Discover the Sky through Binoculars

A systematic guide to Binocular Astronomy

Stephen Tonkin FRAS

BinocularSky Publishing

Discover the Night Sky through Binoculars
A systematic guide to Binocular Astronomy

Copyright © 2018 Stephen Tonkin

ISBN: 978-1-9164850-0-6

First printed October 2018

To the memory of my father, Allan Tonkin, who triggered my astronomical interest in October 1957 when he took me outside to try to see Sputnik, and whose Zeiss 10x50 binoculars were the first instrument I used to observe the night sky.

I am mortal and the creature of a day; yet when I seek the massed wheeling circles of the stars, my feet no longer touch the earth and, side by side with Zeus himself, I sup the nectar of the gods.

Ptolemy of Alexandria

CONTENTS

Contents

Appendices

Acknowledgements

Thanks are due to:

Louise for coffee, moral support and proofreading the manuscript. Any remaining errors are my own responsibility.

The editors of *Sky & Telescope* for permission to use charts from *Sky Atlas 2000.0* (© 1981 by Sky & Telescope, a division of F+W Media, Inc.) as backgrounds for the front cover image and for Figs 2.1 and 2.2

The many fine people that I've met in the astronomy world and who have shared ideas that I can put into use.

Bill Gray of projectpluto.com, whose superb Guide planetarium software has been a valued companion for more than twenty years, and which was used to generate the charts associated with this book.

http://www.messier.seds.org/Pics/More/m1rosse.jpg for Fig.4.2.1 (Public Domain)
https://commons.wikimedia.org/wiki/File:Argo_Navis_Hevelius.jpg for Fig 6.2.1 (Public Domain)
Fig 15.1.1 is Public Domain (reproduction of two-dimensional public domain work of art)

Introduction

The advice given to beginning amateur astronomers has long been "start with binoculars". There are very good reasons for this. Many people already have binoculars or, if not, can obtain reasonable ones very inexpensively, so their initial financial outlay for testing out the new hobby can be minimal. More importantly from an astronomical perspective, the extreme portability of small binoculars, allied with their wide field of view, makes it relatively easy for you to "learn the sky".

But, given the increasing availability of "clever" telescopes using GPS and GOTO technology, whose vendors claim that they are able to get you observing with little or no knowledge of the sky, why should you bother with the old route? The number of appeals for help I get in my capacity as a teacher of astronomy suggests that, clever as it may be, this new technology is far from foolproof, especially where its low-budget incarnations are concerned.

Another phenomenon I've noticed is that, whereas adults tend to want to miss this early step and spend money on flashier kit, youngsters, once given binoculars and a bit of practical advice, thrive on this simple approach. After a recent event our astronomy club hosted for some youngsters, parent reports suggested that the children were as delighted at being able to find a galaxy, scan the

Milky Way, and see double stars for themselves with binoculars as they were at seeing Saturn's rings and Jupiter's moons through a telescope. Often, the flashier kit is consigned to the attic (sometimes to re-emerge on my astronomy courses, accompanied by appeals for help to get it up and running).

This book is primarily for those who want to spend the time building a firm basis in observational astronomy. I have written this primarily as a guide for users of medium-sized (i.e. 10x50 to 15x70 range) binoculars, but it will also be useful for users of small, wide-field telescopes.

If you work your way through this book, by the end of your first year, you will have a sound familiarity with the sky that is visible from the northern hemisphere. You will also have a good idea whether observational astronomy is for you, and whether you really want to plough more money into the seemingly bottomless pit that is our fascinating hobby. If you do delve deeper, what you have learned from this book will serve as a sound basis for your future astronomical endeavours.

A snippet of advice: If your binoculars are smaller than 10x50, don't rush out and buy bigger ones just yet: I have observed most of the targets in this book with 8x42 binoculars, and many of them with 6x30s.

How to use this book

The aim of this book is to familiarise you with the night sky that is visible from temperate northern latitudes.

The first three chapters are, respectively, the "Why, What, and How" of using binoculars for astronomy. The next twelve are month-by-month suggestions of objects to observe, and the book concludes with appendices of stuff you might find useful, as well as an object index.

When you are ready to begin observing, turn to the chapter for the current month. You will see that it is divided into two parts (e.g. July: 10.1 and July: 10.2). Start with the first part (e.g. July: 10.1). All the objects here will be visible with 10x50 binoculars from suburban skies.

Ideally, you will use the book in conjunction with a star atlas. If you do not own one, you can download a set of charts specific to the book from http://binocularsky.com/DNSB_charts.zip. The charts in the book are size-limited by the size of the page, but should just about suffice if you have nothing better to hand; a magnifier would help!

Use the chart in conjunction with the description in the book to see how to find each object. The charts in the book and the downloadable ones are orientated zenith up for approximately 10pm UT (11pm BST) during the month to which they apply. Identify some bright stars in the sky and relate them to the chart. All the instructions for finding the target object begin with either an object you can see with your naked eye or with an object you have already found. Where appropriate, I give directions in relation to another visible object, but they may be given as compass directions. Celestial directions differ slightly from terrestrial ones: north is the direction of the North Celestial Pole (its location in the sky is indicated by the Pole Star), and the apparent rotation of the sky is from east to west. The directions on each chart are indicated in the key at the lower left.

The object description will tell you what you may see. If you are having trouble seeing stuff, go to Chapter 3 (Effective Deployment) and use the techniques it describes for getting the best out of your binoculars. Remember that patience is important, and sometimes spending a while looking at an object enables you to tease more out of it.

Each chapter also contains some information about some aspect of astronomy that may be related to the objects you are observing. I hope you will find it interesting and that it will enhance your enjoyment as well as your general astronomy education.

The second part of each chapter (e.g. July: 10.2) contains a second set of objects. They range quite widely in difficulty and, depending on your experience and your sky conditions, you may need larger binoculars (e.g. 15x70) to see some of them. If you can't see some of them with the binoculars you have, save them for a trip to a dark sky site or, if you get the binocular astronomy "bug", for when you have more experience or larger binoculars.

Distances are given in degrees (e.g. "Pan 6° SE..."). You should know the field of view of your binoculars (it is often stamped on the cover plate). If, say, your binocular has a field of view of 5°, you would interpret the instruction above a "pan a bit more than one field of view to the SE".

Star designations may be given as proper names or as Greek letters (known in astronomy as "Bayer" letters) or numbers ("Flamsteed" numbers) and a constellation abbreviation, e.g. *β CVn* or *31 Psc*. If these are not already familiar to you, they will become so with practice, and this familiarity will aid your communication with other visual observers. Once you have the hang of it, it's much easier than remembering and using Latin constellation genitives, in which the two examples above would be *Beta Canem Venaticorum* and *31 Piscium*. Don't worry about mispronouncing names of constellations or stars, e.g. *Spica*: nearly all of us do it!

α	alpha	β	beta	γ	gamma	δ	delta
ε	epsilon	ζ	zeta	η	eta	θ	theta
ι	iota	κ	kappa	λ	lambda	μ	mu
ν	nu	ξ	xi	o	omicron	π	pi
ρ	rho	σ	sigma	τ	tau	υ	upsilon
φ	phi	χ	chi	ψ	psi	ω	omega

Other abbreviations you will find include:

arcmin: arcminute (1/60°)
arcsec: arcsecond (1/60 arcmin; 1/3600°)
AU: astronomical unit
ly: light-year
mag: magnitude
NSEW: North, South, East, West

Chapter 1 – Two Eyes are Better than One

Why should you bother with binoculars? Surely a telescope will give you a better start in amateur observational astronomy?

There is absolutely no doubt that a small telescope may be better on the Moon, the planets, and some double stars, but there is so much more to see for which a small binocular is very, very much better. You are also, for reasons that will soon become apparent, much more likely to use it and, as the old adage goes, the best equipment is that which you use the most. Even though I own several telescopes, up to 16" (405mm) in aperture, and have been observing the heavens for more than sixty years, my binoculars are still my most used observing equipment.

There are several reasons that binoculars make a good beginners' instrument; here are the most important ones:

1. If you are uncertain whether this hobby is for you, you want to limit your financial outlay. If you don't already have access to suitable binoculars, you can find some very capable binoculars for under £100, and reasonable budget ones for less than half that amount. If you change your mind, your binocular will still be useful for nature

watching, sporting events, and numerous other applications.

2. They are very portable. A typical 10x50 binocular will be about 18 cm (7 in) long and usually weighs a kilogram (2.2 lbs) or less; considerably so in the case of lightweight models. This means they can easily be hand-held, so they do not need to be mounted. You can easily take them anywhere. If you live in a heavily light-polluted area, it is remarkably simple to grab your binoculars and head out to darker skies, whether you walk, cycle, use public transport, or drive. I have a binocular that lives permanently in my car, enabling me to use them whenever celestial – and terrestrial – opportunities present themselves.

3. They are easy to set up. How long does it take to remove a binocular from its case, put the neck-strap over your head and around the back of your neck, take off the lens caps, and lift the binocular to your eyes? If you want to snatch a quick observing session just before bedtime, or if you want to take advantage of a brief gap in the clouds, binoculars could be your answer.

4. Try this experiment. Most of us have one eye that is slightly better than the other. Sit in front of your computer with a document open in front of you and cover the "other" eye, so you are only using your better one. Then gradually reduce the print size on the screen (Usually 'Ctrl -' or 'Ctrl Mousewheel' will do this) until you can just no longer read it. Then uncover your worse eye and note how the text becomes legible again. This is an aspect of the *Binocular Advantage*.

5. Simply put, where observational astronomy is concerned, you can summarise the effect of the *Binocular Advantage* as "approximately x1.4". What this means is that, for faint objects, a binocular will go as deep as a single optic with 1.4 times the aperture, i.e. your 50mm binoculars can potentially show you stars as faint as you can see in a 70mm telescope. Our visual system has evolved using two eyes, so it should not come as a surprise that there is an advantage to using them both. There are two main ways in which using two eyes is beneficial:

    i. Physiological summation: when we use both eyes, we get a better signal to noise ratio, because neural "noise" is partially cancelled. You can easily demonstrate this with a book in low light: wait until you can only just read it with both eyes, then close one of them – it becomes more difficult.

    ii. Statistical summation: for faint objects, the probability of detection is improved if we use two detectors.

Less important reasons include:

6. Each eye has a blind spot, a small region (where the optic nerve enters your retina) that cannot detect light. This is in a different part of the visual field for each eye, so using both of them negates it.

7. Finally, there is false stereopsis: Part of the pleasure of visual astronomy is just enjoying what we see, and although we do not really see distant objects in 3D, it *looks* as if we do. This enhances your pleasure. If you enjoy what

you see, you are more likely to go out and look.

In summary, binoculars offer you an inexpensive route into astronomical observing beyond what you can do with your unaided eye. There are some types of observing, such as estimating the magnitudes of brighter variable stars, at which binoculars excel, and there are many deep sky objects that look very much better in a small astronomical binocular than they do in a similarly priced telescope. It is true that there are aspects of telescope astronomy, such as imaging, that are unavailable to binocular users because binoculars are limited to visual astronomy, but for sheer enjoyment of the night sky they cannot be beaten.

Chapter 2 – Binoculars and BSOs

The short version:

If you don't already have binoculars, get the best you can reasonably afford: nobody has ever complained that the optical or mechanical quality of their kit is too good.

The long version:

Binoculars are specified by a series of numbers and letters, e.g. 10x50 ZCF.GA.WA.WP. The numbers tell you the size of the binocular and the letters give additional information.

The first number is the magnification, the second is the aperture of the objective lens in millimetres. The example above therefore has a magnification ("power") of 10 (your target will appear 10x as large as it does with the unaided eye) and an aperture of 50 mm. If these numbers are correct (and, especially at the budget end of the market, they often aren't) they can give you additional information. For example, the diameter of the exit pupil, which is the size of the beam of light at the point where you should place your eyes, is calculated by aperture/magnification, so a 10x50 will have an exit pupil of 50mm/10 = 5mm. If the exit pupil is larger than your eye's pupil, the effect is that your eye stops down the binocular, reducing its effective aperture. For this reason, it is useful to have an idea of the size of your eyes' dark-adapted pupils. Don't rely on tables, actually measure it (see Appendix D). One advantage of small exit pupils is that they can reduce the effect of aberrations in your eye, such astigmatism.

These numbers can also be used to calculate other factors that you may see in binocular advertising, such as Relative Brightness, Twilight Index, Visibility Factor, and Astro Index. Of these, only the last, which is due to the American inventor and telescope maker, Alan Adler, comes close to the subjective experience of actually comparing binoculars under the night sky. To calculate it, you multiply the magnification by the square root of the aperture, so our 10x50 would have an Astro Index of:

$$10 \times \sqrt{50} = 10 \times 7.1 = 71$$

But remember that, like the other factors, the Astro Index is a rule of thumb, and rules of thumb are merely guidelines, not definitive statements.

The letters tell you about other aspects of the binocular. Our 10x50 ZCF.GA.WA.WP has a body of "Zeiss" (aka "European") construction (Z), is centre-focusing (CF), has rubber armour (GA), wide-angle eyepieces (WA) and is waterproof (WP). There is a more comprehensive list of designation letters in Appendix B.

So what size should you choose? You may hear in the birding community that 8x is the greatest magnification that you can hold steadily. This may be true, but for astronomy we are usually more concerned with detection, and this relates to the apparent brightness of the image we see. Both aperture and magnification aid this.

It is true that no passive optical instrument can make any object brighter than it appears to your unaided eye, so how can a binocular reveal a faint object that you can't see with your naked eye? There are two factors at work, both related to magnification. The first is simply that we find it easier to see things that are larger. The second is that increasing the magnification spreads the light out more and therefore makes any extended object (but not point objects like stars) dimmer. This includes the sky background. The result is higher contrast, so fainter stars are revealed. What you may find surprising is that it also works on extended objects like galaxies, even though they are theoretically dimmed by the same amount as the sky background, so the contrast between the galaxy and the sky background is unchanged. The apparent improvement in contrast is often described as an optical illusion, but it is actually a consequence of our visual system's non-linear response to light intensity.

Fig 2.1 Typical 10x50 and 8x42 centre focus Porro-prism binoculars

Because of this effect of magnification, many amateur astronomers consider the 10x50 as the optimum size for a hand-held astronomical binocular, but there are some that prefer smaller ones, like an 8x42, often because lower magnifications generally give wider fields of view. I have a 6.5x32 that I enjoy because of its lightness and its wide field of view, but my 10x50 is my "go to" instrument at those times when I want to see as much as possible using hand-held binocular. What it boils down to is that binoculars are very personal things so, if you can, try a few before you decide which is best suited to you. Astronomy club observing sessions are ideal for this purpose.

Another popular size for astronomy is 15x70. It is possible to buy very inexpensive binoculars of this size and, whilst they will potentially show you more than a 10x50, even when they are hand-held, they really need to be mounted to make them properly effective. Observing sessions can also help you see what differences there are among binoculars of the same size. These could range from budget

Fig 2.2 Inexpensive 15x70 binoculars

glasses costing less than £20 up to top-of-the-range ones costing around £2000 – yes, you can pay that much for a 10x50! – and, as you would expect, the quality gap is enormous. However, marginal improvements become increasingly expensive, so the image quality difference between £20 binoculars and £150 binoculars will almost certainly be greater than that between the £150 and £2000 binoculars.

Some binoculars do not have a centre-focus wheel, but instead have a separate focuser on each eyepiece. These individual-eyepiece-focusing binoculars are considered to better than centre-focus binoculars for astronomy. This is because they hold focus better and are easier to make waterproof.

Fig 2.3 Individual eyepiece focusing

Unlike terrestrial targets, you don't need to keep refocusing on astronomical targets. It is not necessary, though, and if your binocular will have uses other than astronomy, you may be better off getting a centre-focus binocular, which will be more convenient to refocus on birds, boats or race-horses.

You may also have a choice between roof-prism and Porro-prism binoculars. Most astronomers opt for Porros, simply because in general, at any given price point you get better optical quality than you would with roofs. They are also usually easier to self-maintain.

Three things to watch out for at the budget end are:

1. **Internal stopping.** Often the optical path is stopped down by a fixed iris in the prism housing, so that the 10x50 is effectively about 10x42 and the majority of budget 15x70s are effectively 15x62s. The worst so-called 50mm binocular I've seen is one that was stopped to 10x39. This is done to improve the image quality, at the expense of image brightness. As astronomers, our primary concern is usually brightness.

2. **Double image.** This is almost always a consequence of a prism slipping, but I have seen instances where an objective tube has been cross-threaded and misaligned.

You can't do anything about the first of these, but it is easy to test for (see Appendix E). If you get a double image or a focus problem, return the binocular immediately. Quality control is almost non-existent at the budget end of the market, so a lot of bad beasts get released into the wild. It is also possible that a prism will slip when the binocular is out of warranty; if you want to try to fix that, see http://binocularsky.com/binoc_collimating.php.

If you do want to do some simple tests (mostly useful for comparing binoculars) take a look at Appendix E. There are things that you should definitely avoid. Many of these are characteristics of the BSO in Fig 2.4.

A BSO is a 'binocular-shaped object'. The one shown is among the worst ones that I have had the misfortune to find. Here's a summary of its faults:

Fig 2.4 A binocular-shaped object

1. It is a zoom binocular. There is no such thing as a good zoom binocular. The end.

2. The field of view, especially at the low magnification end of the zoom, is so small it is like looking down a toilet-roll tube (except the image you get from a toilet-roll tube is a lot clearer).

3. It has ruby coatings. Coatings should be anti-reflective, but ruby coatings selectively reflect the red end of the spectrum out of the optical system because the optics are too rubbish to properly deal with chromatic aberration. The result is that the image has a deathly blue-grey cast, so the entire world looks like it's suffered a zombie apocalypse.

4. It does not focus properly; there's a narrow range of "almost there", but no sweet spot, and it's worse away from the

centre of the field of view. The result is the sort of soft-focus effect that is characteristic (or so I am told) of a genre of film that you probably wouldn't take an elderly maiden aunt to see.

5. The eyepiece bridge is so rocky (with the result that it can't even maintain its abysmal best focus condition) that it's a wonder that Sylvester Stallone has not claimed a breach of intellectual property rights!

6. It pretends to be 70mm aperture. This is stopped internally to 50mm, i.e. half the light that struggles through the ruby coated optics doesn't even get as far as the prisms.

7. You do, however, get two images for the price of one!

I am occasionally asked what brand it is. I have no idea; there is no brand name on it. (Well, would you want your name plastered over such an abomination?).

Other things to avoid are "zip-focus" and "focus-free". *Zip focus* is a feature that enables you to change focus from infinity to as close as the binocular will focus with about a quarter of a turn of the focus wheel. This makes it very difficult to attain perfect focus. *Focus free* binoculars are set to what is called the *hyperfocal distance*. This distance, which varies from one optical system to another, is the closest distance at which your eye can compensate and bring all objects beyond that distance to an acceptable focus. They simply aren't very good for astronomy.

Chapter 3 – Effective Deployment

Your first task is to focus your binocular. Choose a moderately bright white star (not a reddish one). The simplest way to point your binocular at it (or any object) is to look at the object with your naked eyes, then just put the binocular in the way. We'll start by focusing the left-hand side. Cap the right-hand objective – don't close your right eye, it will cause the left one to squint a bit – and centre the star in the eyepiece. If you don't have the objective cap available, just put the palm of your hand over the objective lens (without touching the glass, of course). Use the centre-focus wheel (or left eyepiece focuser if you have individually focused eyepieces) to focus the left-hand optical tube, getting the star as sharp as you can. Then uncap the right-hand objective and cap the left hand one and use the right eyepiece dioptre (or right eyepiece focuser) to focus the image in the right eyepiece. If you have centre-focus binoculars, you shouldn't normally need to reset the right eyepiece dioptre adjuster again, and from now on you can focus using the centre-focus wheel only.

The next task is to set the inter-pupillary distance (IPD). This sets the distance between the eyepieces to the distance between your eyes. You can do this on any distant object during daytime or night-time. Adjust the central hinge so that the eyepieces are their maximum distance apart. Look at the object through the binocular and adjust the hinge so that the images from both eyepieces just merge into a single circle. You should not see the sideways figure-8 that you see in films when they are trying to represent a binocular view.

That's it: you are now ready to start observing. If you are completely new to this, spend a few minutes just panning around the sky. Notice how much more you can see through your binoculars as compared to your unaided eyes. Practice locating bright objects without the binoculars, then raising the binoculars to your eyes and getting the object near the centre. If you don't immediately see it in the field of view, you are probably aiming too low. This is a useful skill, and it soon becomes instinctive.

You will also soon get an instinct for how high up you can comfortably observe without straining your neck. For extended periods of time on anything higher than this, you will need to sit or recline, or else observing becomes uncomfortable.

Fig 3.1 10x50 binoculars mounted on a trigger-grip head.

Although one of the reasons for using a 10x50 is that it can be hand-held, it will show you more if you can steady it. Most modern 10x50s have a ¼-inch threaded hole under a removable cover at the objective end of the hinge. This accepts a tripod adaptor, which allows you to mount the binocular on a standard photographic tripod. Whilst tripods are useful for terrestrial viewing and low altitude astronomical viewing, they are a nuisance for anything above about 45°: just too many legs trying to occupy the same space. Monopods are better, and can be more easily used when you are sitting or reclining (they don't have to be perfectly vertical to steady the binocular) but, like tripods, they do reduce the simplicity of observing with handheld binoculars.

If you are using larger binoculars consistently, you may wish to consider a parallelogram mount. They can be home-made if you have the appropriate wood-working or metal-working capability (see links on http://binocularsky.com/binoc_mount.php), or bought at a reasonable price but, for seated/reclined use, do make sure you get one that does not limit you to looking back over the tripod (i.e. you need one with at least 4 "degrees of motion"). Also, if you wish to be able to alternate between reclined and standing use

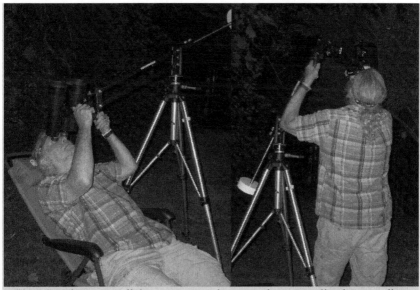

Fig 3.2 A decent parallelogram mount lets you observe reclined or standing

without having to faff about with the tripod height, you need one with parallel arms at least 50 cm (20 inches) long (measured between the pivot centres).

An alternative to a "proper" mount is to stabilise the binoculars with some sort of informal support for your arms, most effectively done at your elbows. You can do this by resting your them on walls, car (automobile) roofs, walls or fences, gates, boulders, rotary washing lines, tables, cushions or other elevated supports on the arm-rests of recliners, the shoulders of a companion... This list is limited only by your imagination and ingenuity. Even if you cannot support your elbows, merely leaning your body against a sturdy object (wall, tree, car, boulder, telegraph pole – or the ground!) will stabilise what you see in your binocular eyepieces. You can also improve the steadiness by changing the way you hold them. The "normal" way with your hands on the prism housings or the objective tubes is fine for targets at the same level as you, but is ergonomically inefficient for high elevations because the higher you raise your arms above the level of your heart, the more tiring it becomes. A simple solution is what is usually called the "triangular arm brace".

Fig 3.3 The "triangular arm brace" hold for steady binoculars

For several decades I have run astronomy clubs for youngsters and we have used 10x50 binoculars as our "standard" instrument. By showing them this way of holding binoculars, I have taught children as young as 10 years old to effectively use these binoculars for observing. If the detail that they report as being able to see is any indication, they are seeing noticeably more than adults using the "normal" hold.

Hold the binocular with your first two fingers around the eyepieces and the other two fingers around the prism housing. Then raise the binocular to your eyes and rest the first knuckle of your thumbs into the indentations on the outside of your eye sockets, so that your hands are held as if you were shielding your eyes from light from the side. Rest the top bone (second phalanx) of your thumb against the outside of the ridge of bone above your eye socket, and the lower bone (first phalanx) against your cheek-bone. Each of your arms is now locked into a stable triangle with your head, neck and shoulder as the third "side", thus giving you a much more stable support for your binoculars. Some of the weight is effectively transferred from your arms to your head and neck, so it is less tiring on your arms, especially if you are reclining. The position of your thumbs keeps the eyepieces a fixed distance from your eyes. You cannot normally reach the focus wheel on centre-focus binoculars when you hold them this way (although you can with roof-prisms), but you should not need to refocus during an observing session. This grip does feel unusual at first, but it is so superior to the "normal" way that it soon becomes second nature.

The best part of the sky, all else being equal, is near the zenith, because that is where we are looking through the least atmosphere. (You have probably noticed that stars near the horizon twinkle more than those near the zenith, for the same reason.) Unfortunately, that is the most uncomfortable part of the sky to look at with hand-held binoculars unless you are lying back. If you are going to get into this astronomy lark, you will eventually want to get some sort of recliner, both for binocular observing and for things like watching meteor showers. There is a lot of choice available, but do avoid the very cheap ones that are sometimes sold in motorway service areas or pound-shops; they do not usually last very long.

It may be that you don't immediately see the target that you are seeking. There are two possible reasons for this: either the object is too faint to be visible against the sky background or it is simply not in the field if view. There's not an awful lot you can do about the first problem other than wait for better conditions and hope. For the second, you need to employ a search strategy (having verified that you are looking in the correct part of the sky). There are two simple methods, each employing what I call a "search-field", which is about half of a field of view. You get some overlap of search scans but, importantly, you are unlikely to miss out any sky and are covering most of it with the "sweet spot" in the middle of your field of view. Once you have experience, vary this to what suits you and your binocular, but half a field of view is a good place to start.

Step search. (Fig 3.4) Use this when you are looking for an object near the horizon. In dawn twilight, first make sure that the Sun is not peeping above your horizon. Start on the horizon, well to one side

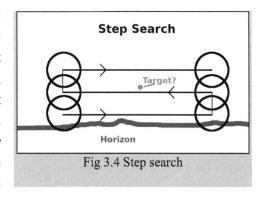

Fig 3.4 Step search

of where you presume the object to be. Scan horizontally until you are the same distance the other side, then raise your binocular by 1 search-field and scan in the other direction. Keep stepping up until you either find the target object or you are looking a couple of search-fields higher than it could possibly be.

Square spiral search. (Fig 3.5) Use this when the target is not visible at the end of a star hop.
1. Scan 1 search-field up.
2. Scan 1 search-field to the right.
3. Scan 2 search-fields down.
4. Scan 2 search-fields to the left.

5. Scan 3 search-fields up.

6. Scan 3 search-fields to the right.

7. … and so on, increasing by one search-field every second scan until you have either found the target object or you are beginning to search sky where it could not possibly be.

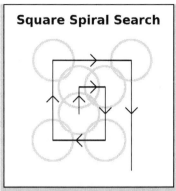

Square Spiral Search

Fig 3.5 Square Spiral Search

Chapter 4.1 – January: The Skies of the Hunter

The January skies are dominated by the distinctive constellation, Orion, which is obvious in the south. Orion is famous for its Great Nebula, M42, object number 42 in Charles Messier's catalogue, and the "Armpit Star", Betelgeuse (α Ori), a red giant that is thought to be close to ending its life in a supernova explosion. It may have already done so, but we would not know until 500 years after the event, which is the time light takes to travel from it to us. When you look at Betelgeuse, consider that if it was where our Sun is, not only would Earth be inside it, but so would Mars and Jupiter.

We start our exploration of the celestial Hunter at the Belt, three stars that easily fit in the field of 10x50 binoculars. The Belt is a good lesson in line of sight illusions: not only is the right-hand star, Mintaka (δ Ori), the faintest, it is also the nearest at just under 700 ly away. The left-hand star, Alnitak (ζ Ori), is only slightly further away, but the brightest central star, Alnilam (ε Ori) is about three times as distant.

The Belt is part of a larger group of stars, an OB association called Collinder (Cr) 70. OB associations are loose groups of between ten

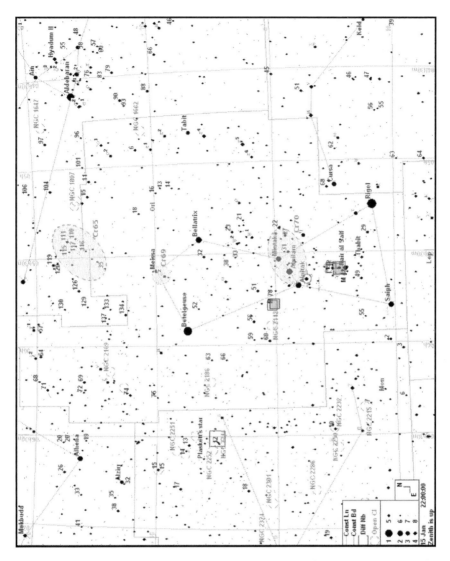

and a hundred young, hot (spectral types O and B), massive stars, together with up to a thousand or more lower mass stars. Your binoculars are ideal for appreciating this group. Note, in particular, the S-shaped chain of stars that weaves between Alnilam and Mintaka; if you allow yourself to see this as the neck of a swan, you may see the rest of this majestic creature extending to Alnitak. As well as the many chains of stars, you should also be able to see some multiple stars, including Mintaka, whose 7th mag companion is 52 arcsec to the N.

Our next target lies about a degree SW of Alnitak. You can see σ Ori with your naked eye as a 4th mag star, but your binoculars should show it to be double. It's a slightly closer double than Mintaka; this time the 7th mag secondary is 42 arcsec to the NE. σ Ori is actually a sextuple star; a decent amateur telescope will tease out the next two components, but you need a very good one and a very high power to see the 5th component. The 6th has never been seen, but its existence is inferred from spectroscopic radial velocity measurements.

Now we head down to the "Sword" and look at the middle "star". This is the famous nebula, M42. It is a fine sight in binoculars of any size, which should also reveal that the other two "stars" of the sword are, in fact, clusters. The one nearer the belt is NGC 1981, and the lower one is NGC 1980. These are both later stages in the evolution of nebulas like M42, which is a star-forming region. The bright nebulosity you see is rarefied gas, and the dark "inlet", often called the "fish's mouth", near the N of it is dust. It is also dust that separates it from the smaller bright patch to the N, M43. The gas and dust are collapsing and condensing to form stars; when it is used up, a small cluster will remain.

Although the nebula looks quite dense, it is actually less dense than the best laboratory vacuum that we can create on Earth! The reason we can see it is that it is so large; it extends for about 20 ly, that is five times the distance from the Sun to the nearest star. Spend as long as you like on the Great Nebula – it rewards patience and the longer you look at it, the more you will see.

When you have finished, make your way to Orion's "head" (λ and φ Ori). Even to your naked eye it will be obvious that this is not a single star, and binoculars reveal it to be a small group. This is another Collinder group, Cr 69, also named the Meissa Cluster; Meissa is the common name of the brightest star in the group (λ). The interest here is the various colours. The original Arabic name of Meissa was Al Hakah, which means "the white spot". Compare this to the sapphire blue and deep yellow-orange of the two φ stars; the yellow one, φ-2, is a foreground star and not actually part of the cluster.

Our last target is yet another Collinder object. Cr 65 is a large asterism that spans about 4° of sky. It is often overlooked because of the more famous objects in this region of the sky. Imagine that Betelgeuse, Meissa, and Bellatrix (γ Ori) form a huge arrow-head. It points to a misty patch on the edge of the Milky Way, about 6.5° from Meissa. The more you look, the more chains and groups of stars you will see. The bright orange star to the N of the group is CE Tau (119 Tau). It is a semi-regular variable star that is similar to Betelgeuse, but is so much fainter because it is both smaller and more than three times as far away.

Chapter 4.2 – January: Starfish and Minnows

Our first object in this set is the lovely "Queen of Clusters", M35. Start at Tejat Posterior (μ Gem), put it on the left-hand edge of your field of view, then look for a large misty patch near the opposite side. M35 looks about the same size as the Moon, and you should be able to resolve at least a dozen stars with 70mm binoculars, even under suburban skies. If you have dark skies, see if you can spot a tiny (5 arcmin diameter) 9th mag cluster ½° towards Propus (1 Gem); you may need averted vision (see Chapter 7.2). This is NGC 2158.

Our next target is an easy asterism which we can find a little more than 4° from Hassaleh (ι Aur) in the direction of θ Aur. Here you will find a little elongated group of 5th mag stars, 14, 16, 17, 18 and 19 Aur, extending SW-NE across a degree of sky. It will be easily visible to your naked eye in a dark sky. Binoculars, allied with a hefty chunk of imagination, will reveal that this group of stars forms the shape of a fish: the Leaping Minnow. If you look to the NE of the Minnow you will see another group of stars, the "Splash", surrounding φ Aur. You should expect to see at least forty stars in the Minnow and Splash combined.

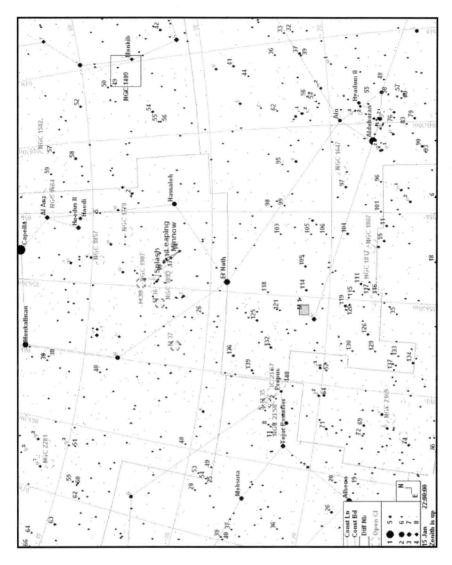

Put φ Aur at the bottom of the field of view, you should see two misty patches, to the left and right of the centre respectively. The right hand, larger and fainter one is the Starfish cluster, M38 and the other one is M36, which is very similar in size and age to the Pleiades but, at 4100 ly, is ten times further than the Pleiades' 403 ly. If you now put M36 at the right-hand edge of the field, near the centre you will see another misty patch, larger and brighter than either M36 or M38. This is M37 which, with a distance of nearly

4,500 ly, is slightly further away than the other two.

We complete our group of watery objects with the one that inadvertently started an observing challenge. In 1705, Edmond Halley predicted that a comet that he had systematically observed in 1682, would return in 1758. On August 28 of that year, King Louis XV's "ferret of comets", Charles Messier was ensconced in Paris's Hotel de Cluny, seeking the comet whose return Halley had predicted half a century earlier. On that night, he saw something that he thought might be a comet, a tiny (5 arcmin diameter) misty patch 1.1° NW of ζ Tau, unaware that it had been observed there some 27 years earlier by John Bevis. A week later, the patch hadn't moved, and he realised that it was not the comet. Thus M1 (Messier 1) became the first in his catalogue of objects that comet-hunters should shun. There are 110 Messier objects, and the challenge, which is only possible for a few weeks in March and April, and ideally from a latitude between 15° N and 35°N, is to observe them all between dusk and dawn. This is possible because they are not evenly distributed across the sky.

Eighty years later, William Parsons (Lord Rosse) observed it through a 36-inch telescope and produced a drawing of it that he thought looked like a horseshoe crab. The later drawing from his 72-inch telescope looked entirely different, but his name "Crab Nebula", had already stuck. By 1913, photographs taken several years apart revealed that it was expanding and in 1939 Nicholas Mayall demonstrated that it derived from the "guest

Fig 4.2.1 Lord Rosse's Crab

star" that Chinese astronomers had observed, in daylight, in 1054. We now know that the "guest star" was a supernova that occurred in the Perseus arm of the Milky Way galaxy. It's not an easy object, though: you'll need a dark, transparent sky to observe this mag 8.4 remnant of SN1054.

Chapter 5.1 – February: Canine Treats

Once you start using them, your binoculars will reveal a wide range of star colours. The colour is related to the temperature, in the same way that a white-hot object is hotter than a red-hot one. In general, stars are hot and blue at the beginning of their lives and (relatively) cool and red towards the ends of their lives. There are exceptions: for example, "red dwarf" stars may never have been very hot. You won't see many of these with binoculars because they are not very bright. The brightest is Lacaille 8760, in the southern constellation of Microscopium, and it is too faint to be seen with the naked eye. Also, don't assume that, merely because a star is red, it is very old. Newly-formed red dwarves would still be red, and different sized stars evolve at different rates. Small stars can live for trillions of years, stars like the Sun live for about ten billion years, but very massive stars will self-destruct in a supernova after only a few million years.

This month we look at Orion's faithful dogs (Canis Major and Canis Minor), and a mysterious faint unicorn (Monoceros) that seems out of place in this celestial hunting scene.

Just below the brightest star in the night sky, Sirius (α CMa), you will find our first object, the open cluster M41. If you have a dark transparent sky, you won't even need binoculars to see it;

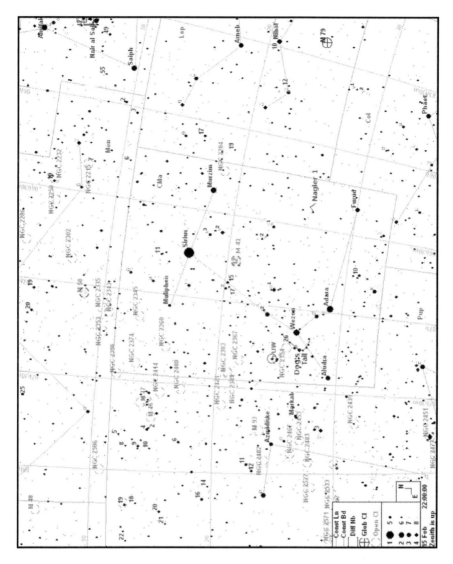

otherwise just put Sirius at the top of your field of view and the cluster will be near the bottom. The earliest known record of it was made by Alexander the Great's famous tutor, Aristotle, in the 4th Century BCE. M41 is a large bright cluster (about the same apparent diameter as the Moon) and, as long as you have a reasonably dark site, you should be able to resolve ten or more brighter stars against the background glow of the fainter members. If your sky is particularly good, you may be able to detect that the brightest star, near the centre of the cluster, is a yellowy orange colour. If you can see this, with patience you may be able to notice some more subtle variations in colour amongst these stars. This variation occurs because the biggest stars here are already evolving into red giants – remember, the more massive a star is, the faster it evolves; a star ten times the mass of the Sun would evolve about a hundred times as fast!

There are some lovely groups of coloured stars in Canis Major and we'll start with the easiest one. Strictly speaking, you don't need binoculars for this group but, as it is low in our skies, they do help to bring out the colours of the stars. The stars we are looking at are Wezen (δ CMa), σ CMa and Adhara (ϵ CMa). The brightest is Adhara, which shines with an intense blue-white light; it is the brightest known source of extreme ultra-violet radiation. It's 430 ly away now, but about 5 million years ago it was only a tenth that distance and would have been about as bright as Venus is today. The next brightest is Wezen, a yellow supergiant, that has stopped fusing hydrogen and is cooling into its red giant phase. In between is orange σ CMa; it's a supernova candidate.

One and a half degrees east of Wezen you will find ω CMa. Note its colour, a brilliant white that you can compare to the fainter yellow-orange star that is just to the S of it. The two are part of the "Dog's Tail" asterism, a semi-circular curve of stars that seems centred on Wezen.

Our next target is yet another remarkably luminous star. UW CMa is 200,000 times as luminous as the Sun, although this is not immediately apparent at a distance of 5,000 ly (although this distance is in dispute). To find it, look about 2.5° NE of Wezen, for the brilliant blue-white τ CMa. UW, which is also a brilliant blue-white, lies a little less than half a degree to the N of τ CMa. UW is a Beta Lyrae type of eclipsing variable. It is eclipsed every 4.39 days by a fainter and smaller companion. This causes its apparent brightness to fall by about half a magnitude as this blue supergiant is eclipsed by its more massive, though fainter ('only' about 60,000 times as bright as the Sun) and smaller companion. It is a contact variable, which means that the stars are so close that their shapes are distorted and matter flows from one to the other.

Our next asterism is Nagler 1. From Phurud (ζ CMa) pan 3.75° in the direction of Mirzam (β CMa). What you are looking for is a single chevron of 7th to 9th mag stars, that looks a bit like a lance-corporal's stripe, and which is pretty much at the limit of visibility in 10x50 binoculars.

Our last object is a lovely open cluster that appears about half the size of the Moon. An easy way to find it is to pan just over 5° NNE from Sirius in the direction of Procyon (α CMi) and look for its very obvious glow. This is M50. Although you may only be able to resolve five or fewer stars, even in very good skies, you are seeing the glow of over 100 stars that span about 20 ly. M50 is 3,200 ly away.

Chapter 5.2 – February: Now You See Me...

There are some faint objects in the second set of this month's targets, so use larger binoculars if you have them. Bigger binoculars will not only show you fainter objects, but they will also bring out more colour in stars. You will also find that averted vision (described in Chapter 7.2) helps you to detect faint objects.

We'll start with NGC 2232 a cluster that can be tricky to identify. About two degrees N of β Mon is the 5th mag star, 10 Mon. Your binoculars will reveal that it is part of a sparse cluster of stars – so sparse, in fact, that you may not, initially, realise that it is a cluster. You will certainly find it difficult to tell where the cluster ends and the surrounding stars begin, but you should resolve 15 or so stars. Notice the narrow vee of 8th and 9th mag stars extending S from 10 Mon, and the slightly brighter (and wonkier) wider vee, that includes 9 Mon, to the N.

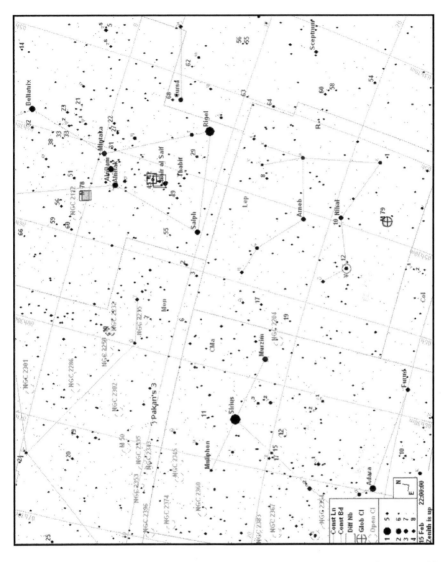

If you don't remember how to find M50, you can remind yourself by looking at the end of the first set of the February objects (Chapter 5.1). About half way between M50 and θ CMa there is a pair of mag 6.5 stars separated by about half a degree. A little more than a degree W of this pair, there is an indistinct "3" shaped group of 9th mag stars. This is Pakan's 3. A lot of sources will tell you that the "3" is reversed; it isn't, at least not when viewed from the northern hemisphere with binoculars.

We're going to hop down to Lepus for our next three targets. The first is the easiest of this set of objects. It is the 4th mag γ Lep, which you can see with your naked eye. This is a double star and even small binoculars will show you the fainter (6th mag) companion, which is a tad over 100 arcsec to the N. γ Lep, which is about three times as bright as the Sun, is quite close to us: a mere 29 ly away. The 6th mag star is AK Lep, which is a BY Draconis type of variable star. The equivalent of huge sunspots pass across its surface as it rotates, causing its brightness to vary by less than a tenth of a magnitude over a period of about 17 days.

M79 is more challenging the further north you are and will probably need averted vision, at least until you get used to what you are looking for. It is a mag 8.6 globular cluster but, even from the south of England, it doesn't get higher than 15° above the horizon, so its light is always extincted (dimmed) by at least half a magnitude by the atmosphere. To find it, start by finding Arneb (α Lep) and Nihal (β Lep) and extend a line from Arneb through Nihal a further 4° S, where you will find a 5th mag star. While you keep your gaze centred on this star, pay attention to the region of sky about half a degree towards γ Lep (the nice easy double we just looked at). Here you should see what looks like an out-of-focus star; it is both fainter and fuzzier than the 5th mag one at the centre. What you are seeing is the combined light of M79's 150,000 stars shining across 42,000 ly of space.

Hind's Crimson Star (R Lep) may be the reddest star in the sky. It is a Mira-type star that varies between magnitudes 10.5 and 5.9 over a period of 445 days, and is near its maximum, and so should be easily visible as it is very well placed for observing in January evenings. Its colour, which is due to an enormous amount of soot (carbon) in its atmosphere, is obviously most apparent when it is near maximum, but is actually redder at its minimum because it is the blowing off of the soot that causes it to brighten. It is from stars like this that most of the carbon on Earth came. To find it, follow the line from α Lep to μ Lep a further 3°, where R Lep should be visible and identifiable by its colour.

You'll want a very transparent sky and may need to mount your binoculars for our last target. Pan 2.5° NNE of Alnitak (ζ Ori) and get that part of sky as near as you can to the centre of the field of view, while keeping Alnitak out of the field of view so you are not affected by its glare. M78 is quite easy to locate as a small misty glow, but merely locating it is not the challenge – you can do that in an 8x40 on a decent night. What we are trying to detect, and the reason we are using larger mounted binoculars, is the brightness gradation. Relax your eyes and look away from it while still concentrating your attention on it (averted vision). As you study it, you will notice that not only is it brighter at the top than it is at the bottom, but the light cut-off at the top is very abrupt: there is a narrow dust-lane here that is responsible. For this reason, it has been mistaken for a comet, and is certainly one of the more comet-like objects in Charles Messier's catalogue of non-comets.

Chapter 6.1 – March: Crustaceans in the Byre

The 88 constellations that we have all grown up with were agreed upon by the International Astronomical Union in 1922, and their boundaries were not established until the Belgian astronomer Eugène Joseph Delporte drew them in 1930. Before this, many different constellations were informally recognised, and various earlier astronomers defined different groups of stars – what we call asterisms – as constellations. One of these was the 16th Century Flemish Astronomer, Petrus Plancius, who defined many constellations that survive to this day, but one that did not make the 1922 cut was Cancer Minor.

To find the Little Crab, start at Pollux (β Gem) and continue a line through κ Gem for a further 6°, where you will find the 5th mag 81 Gem. The Little Crab is a line of 5th and 6th mag stars that extends for nearly 7° from 68 to 85 Gem, and which looks like a slightly longer, but fainter, replica of the constellation Sagitta – and almost nothing like a crab!

Our next object, which was known to the ancient Greeks as Nephelion, the Little Cloud, is one of the best celestial objects for binoculars. You can see M44, also known since 1840 as "The Beehive", with your naked eye as a misty patch, extending for about three times the apparent diameter of the Moon, about 2°

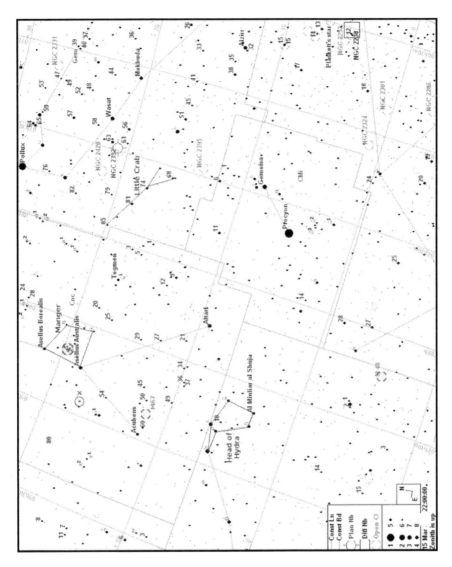

NNE of Asellus Australis (δ Cnc). It appears so large because it is only 577 ly away, making it one of the closest open clusters to Earth, and it is so rich because it has over 1000 stars, more than any other nearby cluster. M44 is in the middle of an asterism called The Manger or The Byre, that has Asellus Australis, Asellus Borealis (γ Cnc), η and θ Cnc at its corners. The Latin for manger, "praesepe", has also been given as a name for the cluster that it surrounds. You may notice that the cluster appears brighter in the middle. This is due to mass segregation, a process by which gravitational interactions between heavy and light stars cause the light ones to move faster and, therefore, further than the more massive ones do in the same period of time. This causes them to end up at greater distances from the cluster centre.

While you're here, pan about 2° east of Asellus Australis, where you will find a short (1°) line of three 6th and 7th mag stars. The deep orange one in the middle is X Cnc, a semiregular variable carbon star which has a mag range of 5.7-6.9 over a period of 180 days. Semiregular variables are massive giant and supergiant stars whose main period underlies irregular changes in magnitude. It has had its apparent diameter measured by lunar occultation. To do this, the time it takes for the star to disappear behind the limb (edge) of the Moon is precisely measured; this gives the diameter as 8.5 milliarcseconds. It is at least 1,000 ly away, so you could use its angular diameter and a bit of simple trigonometry to work out its minimum actual diameter.

Just less than 2° W of Acubens (α Cnc), you will find M67, which has the same apparent diameter as M35 (one of January's objects), although it is not nearly as bright. It appears as a misty patch with 5 stars resolved in 10x50 binoculars. Like M44, it is also brighter in the middle due to mass segregation. It is 4.5 billion years old, which is about the same age as the Solar System. It is also relatively close to us (2,700 ly) and it contains around a hundred Sun-like stars. We believe that it was a cluster just like this that gave birth to our Sun, and, for these reasons, M67 is one of the better studied clusters.

The longest constellation in the sky is Hydra, but only the northern part of it is visible, even from southern British latitudes. The legendary sea-serpent's head is an asterism that is demarcated by six stars that shine between 3rd and 4th mag. We would expect the brightest to be the most likely to have kept their traditional names, but curiously only the faintest of the six, σ Hya, retains a common name, Minliar al-Shuja (the Hydra's nostril), that is related to the Hydra of Greek mythology. Groote Eylandt is, as its name suggests, the largest island in Australia's Gulf of Carpentia. It is owned and inhabited by the Warnindhilyagwa people, for whom this asterism is yet another celestial crab, which they call Unwala. Turn your binoculars onto it and enjoy the wide variation in colours from the yellow-orange of ζ Hya to the intense white of η Hya.

Our final target in this group is sometimes referred to as the "Missing Messier"; it is the open cluster M48. Charles Messier observed it in 1771, but catalogued its position incorrectly (the declination was out by 5°) and nobody could find anything where he said it was. Caroline Herschel re-discovered it in 1783, but it was another century and a half, 1934, before anyone realised that she'd found the object that matched Messier's description. To find it, identify ζ Mon (it is the eastern-most, and brightest, star in an equilateral triangle of 5th-ish mag stars that it makes with 27 and 28 Mon). Now put ζ Mon at the NNW of the field of view, and you should be able to see the 6th mag cluster on the opposite side, 3° to the SSE. It is a condensed patch of stars about the same apparent size as M67. It's quite low for British observers, but in a reasonably good sky you should see ten or more stars against a rich background of fainter ones, with three of the brighter stars forming a tiny right isosceles triangle near the centre of the cluster.

Chapter 6.2 – March: Exploring the Poop

Until the 1922 agreement on constellation definitions Puppis (the Poop Deck) was, along with Carina (the Keel) and Vela (the Sails), part of the huge southern constellation, Argo Navis (the Ship 'Argo' of Jason and the Argonauts fame). Before the ancient Greeks named it, it was the beautiful barque, Neshmet, that transported Osiris in Egyptian mythology. The Bayer letters of all three modern constellations still follow the old Argo Navis designations. This is why the brightest star in Puppis, Naos, is designated ζ Pup and not α Pup - α is given to α Car (Canopus), the brightest star in Carina. Puppis is a Milky Way constellation, so it has lots of open clusters for us to enjoy. Here is a varied selection, but do scan around and see how many more you can find and identify.

Fig 6.2.1 *Argo Navis* from Johann Hevelius's 17th Century *Uranographia*

We'll begin with M93, which is very easy to find and identify. If you place Azmidiske (ξ Pup) at the SE of the field of view, the star-speckled glow of M93 should appear between it and the centre of the field. M93 is a bright (6th mag) cluster, rich with densely packed stars, of which you should be able to see at least two dozen

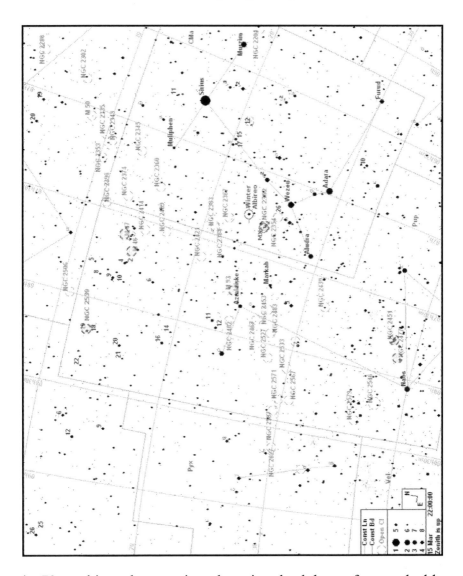

in 70mm binoculars, against the misty backdrop of unresolvable stars. Most open clusters are rich in the middle and sparser at the periphery, but M93 is unusual in that this distribution is reversed. The centre of the cluster, which is marked by a wedge-shaped group of brighter stars, is sparser than the outside.

Find the point where a line going S from α Mon intersects a line going E from Sirius (α CMa). Here, in the same field of view you will see two open clusters that demonstrate how very different they can appear, even though they look approximately the same size. The brighter one is M47 which is closer than the other, M46 (1,600 and 5,500 ly respectively). M47 looks much looser, making it far easier to resolve individual stars: you will get six to ten of them, depending on your sky conditions. The more distant M46 appears much more compact and, like M47, has about 500 stars, but they appear as a diffuse glow and you probably won't be able to resolve any of them out of the background glow. The reason they look the same size is that M46 is much older and has therefore spread out more.

Our next cluster, NGC 2539, is adjacent to the 5th mag 19 Pup, which itself is 8° SE of α Mon. NGC 2539 is a challenging object to locate, but relatively easy to identify. It requires good sky conditions which, owing to its declination, are rare from the latitude of Britain. I find the surrounding star field confusing for star-hopping and my usual method of location is to scan the region with 10x42 binoculars, in which it appears as a faint misty patch, and find the location with these in order to point the larger instrument in the same direction. In the larger instrument it still looks merely as a larger misty patch, but one that is attractive for its delicacy.

NGC 2362 surrounds the 4th mag τ CMa, which is just over one field to the NE of δ CMa. This is a superb cluster in big binoculars, and with 15x70s you should be able to resolve six to ten surrounding the brightest star, τ CMa, which is a brilliant bluish white. It contains several very blue stars, whose presence is an indicator of its comparative youth: it may be less than 5 million years old. Just to the E, the 7th mag star (also bluish-white) is MX CMa. Like the nearby UW CMa (Chapter 5.1), MX is another Beta Lyrae type eclipsing variable. While you are here, you must seek out, 100 arcmin to the N, the beautiful "Winter Albireo", HIP 35210. The orange primary star shines at mag 4.8 and the white secondary, 27 arcsec away, is mag 5.8.

NGC 2451 is 1° N of the centre of a line drawn from ζ Pup to π Pup. It is quite a sparse cluster with several relatively bright stars, which makes it a good object for binocular observation. Adding to the attractiveness of the cluster is the fact that the brightest star is orange and the surrounding bright stars are brilliant white. Six stars are particularly easy to see and a dozen or so are visible in medium binoculars under good conditions. While you are there, see what you can make of the nearby NGC 2477.

Chapter 7.1 – April: Behind the Veil

The "Veil" in the chapter title is that of Thisbē, the lover of Pyramus in Ovid's retelling of the Mesopotamian "star-crossed lovers" myth of Pyramus and Thisbē, which was immortalised by Shakespeare as the "play within a play" in *A Midsummer Night's Dream*. She dropped her veil when fleeing from the lion, leading to the tragic turn of events that led to Pyramus's suicide. In Greek mythology, the same stars are the shorn tresses of Berenice, Queen of Egypt, dedicated to Aphrodite in exchange for her husband's safe return from battle. The constellation name, Coma Berenices, means "Berenice's hair".

If ever a celestial object seems to have been made for binoculars, it is Melotte 111, which you can see with your naked eye as a misty patch between Cor Caroli (α CVn) and Denebola (β Leo). It is obvious how it represented a veil. It covers a diameter of nearly 6°, and so fills the view of small and medium binoculars. Melotte 111 has a paucity of faint stars: there are none fainter than mag, 10.5. This probably because the stars that are there have insufficient mass to gravitationally bind the lighter stars, and so mass segregation has resulted in the lighter ones leaving the cluster altogether.

Next, identify γ Com. Embedded in the cluster, nearly 2.5° S of γ

48

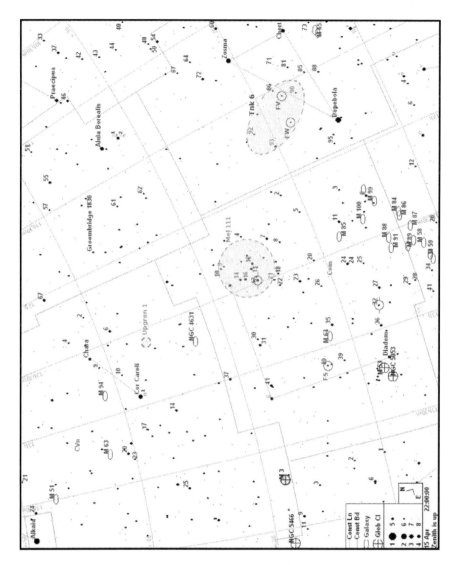

Com, is the double star 17 Com. With a separation of 145 arcsec, the white mag 5.3 and 6.6 components are an easy split in your binoculars.

Find your way to the midpoint between Diadem (α Com) and β Com where, a degree to the W, glowing somewhere around 6th mag, is an orange star, FS Com (also designated 40 Com). It's easy to confirm that you have the correct star by looking for the mag 6.8

star that lies just less than half a degree S. FS Com is a semi-regular variable star whose magnitude oscillates between 6.1 and 5.3 over a period of 58 days. Doppler shifts identified in the spectroscopic analysis of the star's light shows variations in radial velocity. These indicate that its variability is due to pulsations in size.

A bit more than 2.5° W of Diadem you will see an orange star, the 5th mag 36 Com. It is the easternmost of an equally spaced triplet of stars spanning 3° of sky. Our second double star in this month's set of objects is the middle "one", 32 and 33 Com. At 196 arcsec, it is an even easier split than 17 Com, but it is a bit fainter, with the two stars shining at mag 6.3 and 6.9 respectively. Compare the colours to those of 17 Com which we observed earlier: even in small binoculars, 32 Com is distinctly yellowish, but you probably won't be able to detect any colour at all in the fainter 33 Com.

Our next target, M3, is one of the best globular clusters in the northern sky. There are no nearby bright stars to act as pointers, but if you look where a line from Cor Caroli (α CVn) to Arcturus (α Boo) crosses a line drawn between Seginus (γ Boo) and Diadem, you will find what looks like a badly mis-focused star in the field of view. Use averted vision by looking at the bright star ½° to the SW while keeping your attention on the mis-focused star, you should notice that the mis-focused 'star' brightens and grows. This is the glow of the quarter of a million or more stars in the cluster. In his *Celestial Handbook*, Robert Burnham reported that the brightest 45,000 of these stars (i.e. brighter than mag 22.5) were counted manually on a photographic plate at the Mount Palomar observatory! This fine cluster lies 34,000 ly away.

How would you like the opportunity of a very long-term observing project? If you would, Groombridge 1830 is the star with the third greatest proper motion (the apparent motion of a star relative to the Celestial Sphere) and, at mag 6.5, it is both brighter and easier to find than the two that are faster. You will find it almost exactly half way between Chara (β CVn) and Alula Borealis (ν UMa). If you spend fifteen or twenty years very carefully plotting its position in relation to its surrounding stars, you should just begin to notice that it is moving towards the 6th mag red star that is about 1° to the SSE – it'll get there in about 500 years!

The Upgren catalogue consists of exactly one object, and nobody is quite sure what it is: is Upgren 1 merely an asterism, is it a very old open cluster, or is it a chance line-of-sight pairing of two open clusters? To find it, take a line between Cor Caroli and Chara as being one side of an equilateral triangle, and go to its third apex to the SW. Slightly closer to Cor Caroli, you should see a chair-shaped group of seven 8th and 9th mag stars in a region about 30 x 15 arcmin. We don't need to know exactly what sort of object it is to be able to appreciate that, in binoculars, it is a very pretty group of stars.

Binoculars are the ideal instrument for enjoying pretty star fields, so let's head over to one of the finest, Tonkin 6. Find 86 Leo (6th mag, deep yellow) mid-way between Denebola and Zosma (δ Leo). First of all, examine the curved string of 7th and 8th mag stars that stretches 3° eastward. The nearest one to 86 Leo is, an orange long-period variable star, FV Leo and the furthest is FW Leo, which is blue-white. Then return to 86 Leo, and compare its colour to that of the hot blue-white 90 Leo that is 2° away in the direction of Denebola. Now you're getting your eye in, spend some time finding chains and loops of star, and look for the subtle colour differences between them. 92 Leo and 93 Leo, which frame the NE boundary of this star field, serve as a fine example of subtlety in this regard.

Chapter 7.2 – April: Look Aside, Now

Sometimes we need to enhance the visibility of an object merely to make it visible. One way of doing this is to employ a technique called "averted vision", or AV. This makes use of the fact that our eyes have two different types of light receptor in the retina: cones and rods. The cones are responsible for colour vision and are primarily clustered in and around a region called the fovea, which is directly behind the pupil. However, they are not as light-sensitive as the rods; this is why colours seem to fade to shades of grey during twilight. The rods, which are suited for scotopic (dark adapted) vision and are found away from the fovea, are not sensitive to different colours. To use averted vision, we direct our gaze away from the target so that its light falls on the rods, whilst still concentrating our attention on the target. We've already used it a bit, but we'll play with it a bit a bit more to really practice it.

First of all, locate Diadem (α Com) and navigate 1° NE of it to a small misty patch. This is the mag 7.7 globular cluster, M53. Once you have identified it, centre it in the field of view but centre your gaze on Diadem, while still concentrating your attention on the cluster. Did you notice how it seemed to brighten and grow when you looked away? This is the result of averted vision and the phenomena you witnessed is typical of globular clusters, although not all show it to as pronounced a degree as M53. Return your gaze

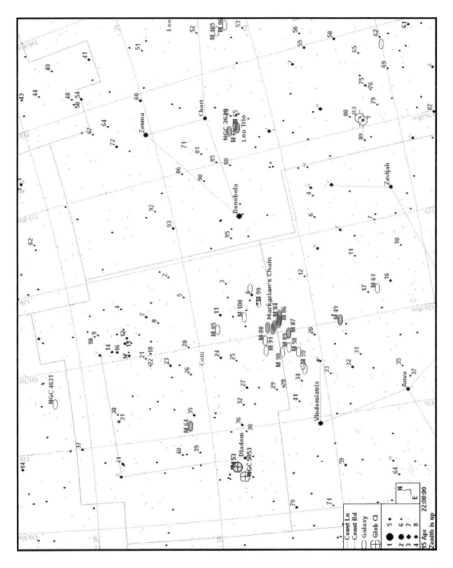

to M53 and avert it to somewhere else in the field of view; experiment to find where to look to get the best difference between direct and averted vision. Spend a bit of time practicing this technique; you'll definitely need it a bit later on when we start looking for faint galaxies in Markarian's Chain.

One third of the way from Diadem to γ Com lies another misty patch, the first of our galaxies, the mag 8.5 M64. It's quite easy to see in the right conditions – which, for all but the brightest galaxies, means a transparent, moonless sky away from sources of light pollution – but do not expect binoculars to show you the dark dust lane that gives it its common name, the Black-Eye Galaxy. All you will see is a small oval glow that extends for about a quarter of the apparent diameter of the Moon, but any Moon or other light pollution will reduce its apparent size quite dramatically. It is 24 million ly away, so the light you are looking at left there round about the time the ape and monkey lineages became distinct and separated from our common ancestor, and Earth's first grasslands were evolving.

The reason that we are looking for galaxies during these spring evenings is that the meridian lies away from the Milky Way, letting us look away from the plane of our galaxy and its obscuring dust, enabling us to more easily see the galaxies that lie in deep space. The first we'll look at is the giant elliptical galaxy M87, which dominates a huge group of galaxies known as the Virgo-Coma galaxy cluster. If you look mid-way between Vindemiatrix (ε Vir) and Denebola (β Leo), you'll see M87 just below left of the centre of the field of view. It appears brighter and a bit smaller than M64, and is circular so you could mistake it for a globular cluster, but a tell-tale difference is that it looks almost uniformly bright. You should be able to see more galaxies in the same field of view. Do you notice how some of them disappear when you look directly at them? The effects of direct and averted vision again. You may see M87 described as a "radio galaxy". This means that it is very luminous in radio wavelengths. They are important to cosmologists because they are detectable at enormous distances.

Our next galaxy, M49, is another elliptical radio galaxy. Place ρ Vir at the NE (top left) on the NE of the field of view; on the opposite side of the field is a pair of orange-yellow 6th mag stars that are separated by just over a degree. M49 is mid-way between them, pretty much the same size and brightness as M64, which you observed earlier, but more circular – like M87, it could be mistaken for a globular cluster. You will almost certainly need a combination of averted vision and dark transparent skies to detect its very condensed nucleus, which can appear stellar in binoculars.

Markarian's Chain is one of the binocular delights of Spring but, to get the best out of it, you will need a very dark transparent sky away from light pollution. It lies almost exactly half way between Vindemiatrix and Denebola. You shouldn't have any difficulty finding galaxies in this region of sky, but identifying them can be much more of a challenge. This is because it can be quite tricky with binoculars to identify the ones that fade in and out of visibility as you change between averted and direct vision. One useful way of recording them is drawing, but be warned: once you start, you may spend an hour or more and still not sketch all the ones you can see, especially if it is a good observing night! Start with the brightest ones, M84 and M86, both of which are around 9th mag, Once you have nailed those down, with a bit of patience you should be able to identify at least the six or seven brightest galaxies of the chain.

Like Markarian's Chain, the "Leo Trio" (M65, M66, NGC 3628) can be a challenge in anything other than a dark, very transparent sky, so wait until Leo is near its highest. This is around 22:00 UT in mid-April, an hour later or earlier at the beginning and end of the month respectively. Find Chort (θ Leo) and pan to the SE until it is just outside the field of view of 15x70 binoculars. The galaxy group will be near the centre. Use averted vision if you need to, but with a bit of practice and familiarity you'll find that they become easier to find. Once you know what you are looking at, you can play: look directly at one of them, and notice how the other two not only become more distinct, but NGC 3628 becomes distinctly more elongated if you are looking at one of the others. If you have mounted binoculars, this is a nice way of demonstrating averted vision to people who have not used it before.

Lyra has a famous "double double" star (ε Lyr), but did you know that Leo has one as well? And that it's a much easier one? If you pan 6° WNW of Zavijah (β Vir), the 5th mag τ Leo is the brightest star in your field of view. Its companion, 83 Leo, is a magnitude and a half less bright and 20 arcmin to its NW. If you look carefully, you will see that each of these stars is itself a double star, with both companions being around mag 7.5, a magnitude fainter than 83 Leo, but still easily within range of binoculars. τ Leo's pale-yellow companion is much easier to split at 89 arcsec from the primary star, but 83 Leo's yellow companion is only 28 arcsec away, making it a bit more of a challenge – make sure that your binocular is well focused; mounting it will also help.

Chapter 8.1 – May: A Royal Visit

The royals are King Cepheus and Queen Cassiopeia, the mythological parents of Andromeda. This circumpolar region around the northern Milky Way is a richly rewarding part of the sky to explore. Not only is it brimming with open clusters, as you would expect, but the sheer number of stars makes some degree of pareidolia (which we examine in more detail in November) almost inevitable.

The first of these is one with which the west-country amateur astronomer, Eddie Carpenter, has for decades been delighting anyone who would care to look. Eddie's Coaster, as it is now known, is an asterism that is not immediately apparent in images or on star charts but, for some reason, it is immediately very obvious in 10x50 or 8x42 binoculars under a decent suburban sky. About 3° N of γ Cas there is a double-peaked wave of 7th and 8th mag stars, reminiscent of a roller-coaster, hence its name.

Imagine a line from Segin (ε Cas) to Ruchbah (δ Cas) as being the SE base of an equilateral triangle. At the 3rd apex is a mag 5.5 star (32 Cas). 1° back towards Segin is 35 Cas, which is nearly a magnitude fainter at mag 6.3, and its mag 8.4 companion, nearly an arcmin to the N, is very easy to distinguish in 10x50 binoculars. The primary star of this optical (i.e. not a true binary star) double is

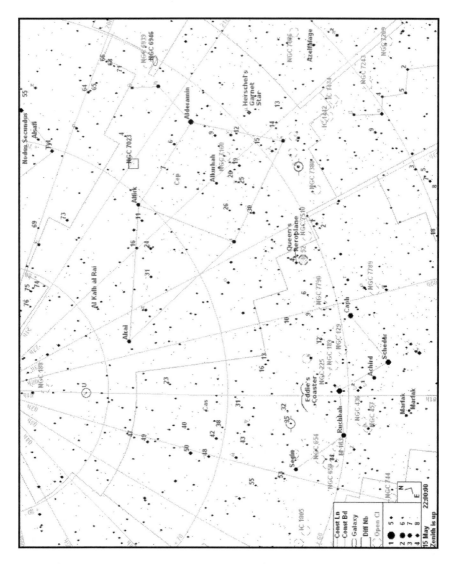

very white, a lovely contrast with the ruddy companion.

The 7th mag M52 is in a straight line with Shedar and Caph (α and β Cas), 6° beyond Caph. It lies within an isosceles triangle of stars that has two 7th mag stars as its base and the golden-yellow 5th mag 4 Cas as its acute N apex. You may see up to ten resolved stars, the brightest of which is an 8th mag one on the western side, against a wedge-shaped grainy glow that is ¼° long. The grainy glow itself, which is easy to distinguish from the Milky Way, is due to nearly 1,000 stars. It is 4,600 ly away and extends over about 18 ly. It was discovered in 1774 by Charles Messier, who happened upon it by chance when he was looking for Comet Montaigne, which was passing nearby.

We just used 4 Cas to help locate M52. Look at it again, and see that it marks the starboard wingtip (yes, I know that it is the port wingtip that should have the red light!) of our next object, a pretty aeroplane-shaped asterism. The stars that define the shape of the Queen's Aeroplane are 8th mag or brighter, so it is easy to identify, even in moderately light-polluted skies, but you will need dark transparent skies if you are to appreciate the wide variety of colours in these stars. The port wingtip is 1° W of 4 Cas; it is a 7th mag brilliant white star, V649 Cas. The tip of the tail is a 7th mag orange-red star that is nearly 1° N of 4 Cas.

δ Cep is the star that gave its name to an entire class of variable stars, the Cepheid Variables. It was not the first to be discovered, though. That honour goes to η Aql, which was discovered a few months earlier. Henrietta Leavitt demonstrated that there is a simple relationship between the period of variation and the luminosity of Cepheids, so merely by timing its period of variability, you can establish its intrinsic brightness. This relationship, which was established by her study of 1777 similar stars, allowed them to be used as the first "standard candle" when astronomers were trying to reckon the size of the Universe. δ Cep varies from mag 3.6 to 4.5 over a period of 5.37 days. But that is only part of its attraction: it is the primary component of a lovely double star. δ Cep itself is a golden colour, and the 6th mag secondary is blue-white. They are separated by 41 arcsec, so are easy to split with hand-held 10x50 binoculars.

Just S of the mid-point of a line between Alderamin (α Cep) and ζ Cep is a distinctly red-orange 4th mag star, μ Cep, named for William Herschel, who observed it in 1783. Herschel's description of how to appreciate the star is as apt today as it was when it was written: "of a very fine deep garnet colour and ... a most beautiful object, especially if we look for some time at a white star before we turn ... to it, such as Alpha Cephei, which is near at hand." This red supergiant is one of the largest known stars: if our Sun was the same size, its surface would be between the orbits of Jupiter and Saturn!

Cepheus is the second most northerly constellation after Ursa Minor, and it is to the N part of it that we turn for our last target in this set. U Cep is an eclipsing variable star. This means that it has a fainter companion that periodically passes between us and it, attenuating its light. Come down 7.5° from Polaris, in the direction of γ Cas and you'll find U Cep in between two white 6th mag stars. It has a nine-fold variation in brightness (mag 6.8 to 9.2) and, being circumpolar from Britain, this makes it a very suitable for newcomers to variable star observing with binoculars. Its period of 2.49 days means that, if you observe an eclipse, you will be able to observe another one 5 days later at approximately the same time.

Chapter 8.2 – May: Island Universes

Only a century ago, it was unclear whether or not some "nebulas", especially the "spiral nebulas", were a similar distance to the stars, or if they were "island universes" at far greater distances. Now, of course, we know that the spiral nebulas – galaxies, as we now call them – are hundreds of times more distant than the most distant stars we can see with our binoculars. Along with some other interesting objects in the region, we'll look at some "island universes" that lie in and around the "Big Dipper" asterism of UMa.

We'll start with the brightest objects in this set, the galaxy pair M81 (Bode's Nebula) and M82 (The Cigar Galaxy). They are very easy to find: merely travel along a line through Phecda (γ UMa) and Dubhe (α UMa) and continue it the same distance to the NW. The galaxies will be in the same field of view, and just N of, the end of this line. M81 is bigger, and is easily the brighter of the pair (mag 7.0, as compared to M82's mag 8.6). Like the galaxies we looked at in Leo last month, you can use these galaxies to demonstrate the use of averted vision: as you direct your gaze at one, the other appears bigger, brighter and more detailed. At 18 million ly distant, M81 is two and a half times as far away as the fainter M82.

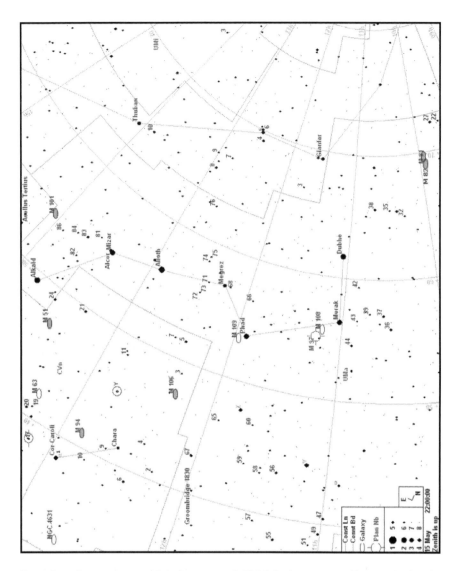

People often assume that because M101 is has a smaller magnitude (7.6) than M82, it will be easier to see. This is not the case. Although it has a greater integrated brightness, it covers a large amount of sky, nearly half a degree across. The consequence of its light being spread over a lot of sky is that its actual surface brightness is very low. This makes it quite a difficult object and, to maximise your chance of success, you should use mounted binoculars under a moonless transparent sky, and make sure that the galaxy is well above any horizon skyglow. If you imagine an equilateral triangle with Mizar (ζ UMa) and Alkaid (η UMa) as its base; M101 is just inside the third apex. If you can't see it at first, use averted vision and try giving the side of the binoculars a sharp tap with your fingertips to set up a short vibration. What you are looking for is a small patch of sky that is very slightly brighter than the rest: this is your galaxy. Once you know what you are looking for, it will become much easier to locate. If you think it is difficult in binoculars, remember that it is one of the objects that is far easier in budget binoculars than it is in an equivalently priced small telescope!

The famous "Whirlpool Galaxy" of Lord Rosse is much easier to find. Take a line from Mizar to Alkaid and, at Alkaid, take a right angle to the SW and pan 3.5° from Alkaid. M51 is a small misty patch that matches Charles Messier's original description, "a very faint nebula, without stars". With small binoculars, you won't be able to make out any structure beyond a slightly elongated appearance and, if the sky conditions are good, possibly the two brighter nuclei of what is actually a pair of interacting galaxies.

Our next island universe, M94, is slightly easier to find. Identify Cor Caroli (α CVn) and Chara (β CVn) and find the mid-point between them. From here, offset 1.5° towards Alkaid – for future reference, M94, M51 and Alkaid are in a straight line. Here, possibly needing averted vision, you should find the fuzzy patch of light that has taken 13.6 million years to reach you from this 9th mag galaxy. You should find it quite easy with 70mm binoculars under dark suburban skies.

Return to Cor Caroli and go slightly more than 2½° E. Our first non-galactic target is a nice easy (4.6 arcmin separation) visual double star, 15 and 17 CVn; they shine at mags 6.3 and 5.9

respectively. This optical double star (i.e. they are not physically related) is a good illustration of how the inverse square law relates distance and magnitude. Although they appear to us to be of similar brightness, 15 CVn is 1,200 ly away and around 360 times as luminous as the Sun, but 17 CVn is about 4% as luminous (but still nearly 15 times as luminous as the Sun!) and just over 16% of the distance to 15 CVn.

Our next target is another star. This time start at Chara and pan 4½° (a little more than a 15x70 field of view) towards Mizar. You are seeking a pale orange star. This is La Superba (Y CVn), a cool carbon star whose magnitude varies from 6.3 to 4.7 and back over a period of about 160 days. Given its common name, you would be forgiven if you expected a more impressive sight, but it was not named for anything you can detect in a binocular. What is "superb" about is its unusual spectrum, in which the blue-violet end is severely dimmed by absorption from organic compounds in its atmosphere.

We'll finish off with another galaxy, M106, which is half way between Phecda (aka Phad) (γ UMa) and Chara. Alternatively, follow the chain of 8th mag stars to the E of Phecda until you get to the obviously brighter (5th mag) 5 CVn, then head S until you get to the equally bright 3 CVn, which is just over half a degree N of the galaxy. This mag 9.1 galaxy is an easy object in binoculars of 70mm or greater aperture. It is elongated in an SE-NW orientation; this is noticeable (possibly needing averted vision) in a 15x70, and is obvious in anything bigger, where it appears as a bright misty glow around a slightly oval nucleus.

Chapter 9.1 – June: Wrestling with the Snake

The southern Milky Way is a lovely part of the sky, marred only by the fact that it's never far above the horizon from Britain. Because we are looking in the direction of the galactic core, it is potentially much denser in that direction than it is in the winter when we are looking away from the core out through the spiral arms. The "confounding" phenomenon is that, as well as there being many more stars, there is also much more obscuring dust. The bulk of visible objects here are, as we would expect, open clusters, amongst the best class of object for binoculars.

I find that it's very satisfying to begin an evening's observing with an easy object that seems to have been created for binocular observers. Melotte 186, an open cluster in Ophiuchus, is a prime example such an object. It has a diameter of about 4° and so fits comfortably in the field of view 10x50 binoculars. The brightest star is the 4th mag 67 Oph, in the centre of the cluster. Other bright stars include 66, 67, 68, 70 and 73 Oph, which form a vee shape that recalls of the Hyades cluster in Taurus. This was the reason that the group was called Taurus Poniatovii (Poniatowski's Bull), named in the 18th Century by a Lithuanian astronomer, Marcin Odlanicki Poczobutt for Stanislaus Poniatowski, king of Poland, who was his patron.

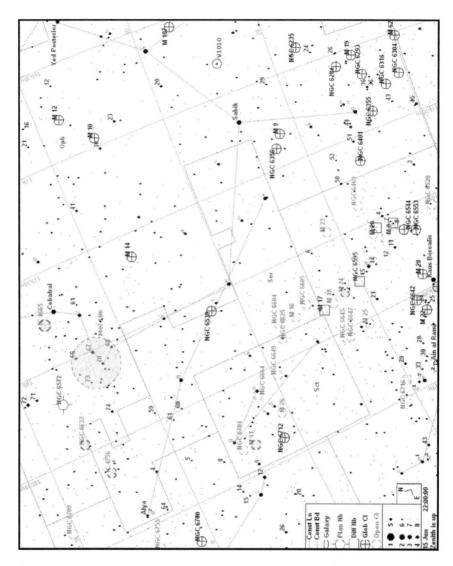

Next, find Cebelrai (β Oph) and, in the same field of view, see the beautiful Summer Beehive, IC 4665, just to the NE. If you look carefully, you'll see that the brightest stars form the upside-down word "HI", a stellar welcome to these summer skies. Your 10x50s should be able to resolve at least a dozen stars in this cluster, which is a lovely sight for any size of binocular. Can you see the curved chain of hot white stars that forms part of the letter "H"? The Summer Beehive, which lies 1,400 ly away, is less than 40 million years old, which is relatively young for an open cluster.

If you go back to Mel 186 and pan about 7° to the NE, you will find NGC 6633. This marvellous cluster can be tricky to find against the background of Milky Way stars, and so is often overlooked in lists of binocular objects but, once you have it, it is easily to see in your 10x50s. You should be able to see the four brightest stars shining against the 20-arcmin elongated glow of the unresolved fainter cluster stars. You may notice that its stars are not as brilliant white as those of the Summer Beehive. They are yellower because they are older. Yet NGC 6633 is not a particularly old object; with an age of about 700 million years, this 1,040 ly distant cluster is merely middle-aged.

Identify the 5th mag Alya (θ Ser) and find the spot half way between it and NGC 6633, where you will find a softly glowing patch just over 1° in diameter. This is IC 4756, aka Graff's Cluster. This huge grouping of stars is over 20 ly across and approximately 1,300 ly away. Be patient and try averted vision on it when, although it can merge with the background Milky Way, it should reward you with far more detail and you may understand why it has been described as "brighter stars, scattered over a background of diamond dust". Its yellowish stars suggest that it is a similar age to NGC 6633, with which it makes a lovely contrast. It is named for Kasimir Graff, director of the Vienna Observatory, who discovered it independently in 1922 (he was unaware that it had been discovered by the American Solon Bailey some years previously).

The densest known open cluster is M11, the Wild Duck Cluster. Were it not so rich, it would be very difficult to distinguish from the background stars, which themselves form one of the most densely-packed regions of the Milky Way, the Scutum Star Cloud.

To find M11, look 2° to the SE of the 4th mag β Sct. In your 10x50 binoculars you will see a bright, slightly wedge-shaped glow, spanning about ¼° of sky. You would be forgiven if you initially thought it might be a globular cluster: at low magnifications, it looks just like one. Your binoculars don't magnify enough to show you the vee-shape of brighter stars that give this cluster its common name, but it is still one of the better objects for binoculars. You should also try panning around the Scutum Star Cloud, whose densely packed stars make it a delightful binocular object in its own right.

Our last object in this set is a rapidly varying star. V1010 Oph changes magnitude between 6.1 and 6.9 over a period of slightly less than 15 hours, but its fall in brightness and subsequent return to maximum takes only four hours This means that you can observe a magnitude change during a single observing session. V1010 is almost certainly a contact binary star (a double star system whose members are so close that their gas envelopes merge with each other) and what we are witnessing is a partial eclipse due to the fainter member passing in front of its brighter companion. To find this periodic variable, start at Sabik (η Oph) and go 5° W where you will find a pair of 6th mag stars less than 1° apart. The more southerly one is our target.

Chapter 9.2 – June: Hanging In The Balance

Our next set of objects lie in and around the constellation Libra. This is an anomaly amongst the zodiacal constellations. Firstly, it is the only one that represents an inanimate object, the Scales. Zodiac (which means "circle of animals") and zoology have the same etymological root. When we look two of my favourite star names, we get a hint at the reason for the anomalous nature of Libra: the names Zubenelschemali (β Lib) and Zubenelgenubi (α-2 Lib) mean, respectively, the "northern claw" and the "southern claw". If you look at the pictorial representation of the constellation directly east of Libra, Scorpius, you may note that it has tiny claws for a scorpion. The region of the constellation Libra was once known as Chelae, "the claws", and Ptolemy of Alexandria (100-170 CE) referred to Zubenelschemali and Zubenelgenubi as "the points of the claws of Scorpio". It appears that some time before the 10th Century CE, the claws of this mythological scorpion were surgically removed to make an additional constellation in order to force-fit a preconceived mystical notion that there should be twelve zodiacal constellations.

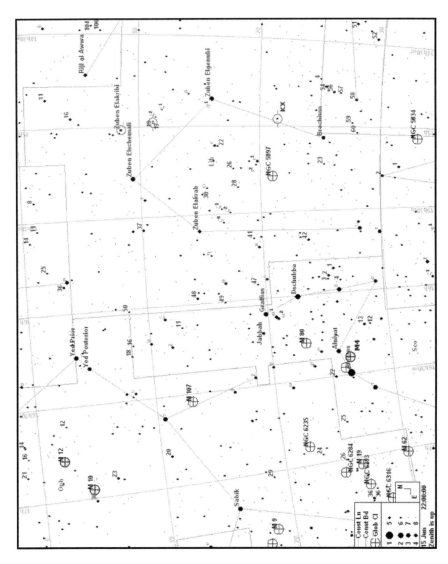

And, while we are on the subject of the zodiac, which is nominally the region of the sky through which the Sun, Moon and naked-eye planets pass, it is worth noting that not only are there twenty-five such constellations, but the Sun spends three times as long passing through Ophiuchus, which is not reckoned to be a zodiacal constellation, than through Scorpius, which is counted amongst them.

Our first stop is the variable (mag 5.8 to 4.4) Zubenelakribi (δ Lib), which you will find 4° W of Zubenelschemali. Zubenelakribi is an eclipsing binary star. What this means is that we have a pair of stars that are in orbit around their mutual centre of mass, and that their orbital plane is edge-on to us so the stars, from our perspective, pass in front of each other. When the larger and dimmer of the two passes in front of the other, occulting it, their combined brightness falls over a period of about six hours. Zubenelakribi has a period of 2.3 days so, if you observe one eclipse, the next will be in two days' time, but nearly eight hours later. Perhaps easier to remember, is that there will be an eclipse 1 week later about half an hour earlier. This means that, even with the short June nights you should have a few opportunities to see an eclipse event and, of course, you can keep observing it outside June!

About mid-way between Zubenelakribi and Zubenelgenubi there are two very easy pairs of double stars. The very wide pair of orange stars are ξ-1 and ξ-2 Lib. They are 0.75° apart and shine at mag 5.8 and 5.4 respectively. Pan 0.5° NE of ξ-2 Lib to find 17 and 18 Lib. These are a bit dimmer at mag 6.6 and 5.8 respectively and are closer at 10 arcmin apart, but unlike the ξ pair, they are contrasting colours. Neither of these doubles is a binary pair where the stars are gravitationally related; they are optical doubles and their apparent closeness is merely a line of sight illusion. 17 Lib is 50 ly further away than 18 Lib, but ξ-2 Lib is more than 200 ly further than ξ-1 Lib.

Identify Brachium (σ Lib) and pan 4° towards, where you will find a yellowish 6th mag star. A closer inspection will show you that this is actually a double star, with an 8th mag companion 26 arcsec to the NW. Unlike the previous double stars, this one is a binary star. The brighter of these is KX Lib. The KX indicates that it is a variable star, but don't expect to be able to detect the variability through your binoculars: it's less than a twentieth of a magnitude. KX is what is known as a BY Draconis star. This means that its variability is due to the sort of activity that we see on the Sun – sunspots and flares – and the star's brightness changes as these move across its surface as it rotates.

We leave Libra and head into Ophiuchus for our next target, the 7th mag globular cluster M12. Imagine that Yed Prior (δ Oph) and ζ Oph are the base angles of an equilateral triangle; you will find M12 at the 3rd (NE) apex. M12 is one of the larger globular clusters in the southern sky. It is about 10 arcmin across, which translates to an actual diameter of about 75ly. For some time, astronomers thought that it might be a tight open cluster, because it does not have a very distinct core. As our understanding of the dynamics of clusters has developed, this has changed, and it is now thought that the combined gravity of the Milky Way galaxy has stripped off many of M12's low-mass stars. Let me take this opportunity to remind you that, like all globular clusters, it benefits from averted vision, with which it appears to both brighten and enlarge.

The easiest way to find our next target, M10 is to start from M12. Place M12 near the NW of (top right at this time of year) the field of view, and M10 will be just over 3° away, near the SE (bottom) side of the field. It looks the same size as M12 but is about half a magnitude dimmer. If you mount your binoculars and use averted vision you should notice that M10 has a more distinct brightening of the core than M12. An easy way to do this is to put both M10 and M12 in the field of view and, when you direct your gaze to one of them, you should notice the other one grows and brightens. The bright star to the east of M10 in the same field of view is the 5th mag 30 Oph; if your sky is dark enough to see it with your naked eye, you can use this star as an easy signpost to the location of the cluster.

We'll finish with what is nominally the brightest globular cluster of the three that we shall view this month. Start at Antares (α Sco) and look just over a degree to the W. That fuzzy blob is M4. It is quite close to us at only 7,000 ly. Owing to this proximity, it appears as a rather loose cluster and is one of few in which some detail is apparent in a 15x70 or 20x80 binocular. It would be even more spectacular were it not for the intervening dust that sculpts the dark lanes of the Milky Way. Because of its location near the plane of the Milky Way, it is in a beautifully rich star-field that is more pleasing in binoculars than in a telescope. It is the one of the largest globular clusters that is visible from the latitude of Britain, but is so low on the horizon that it is less easy to observe than M13 (one of next month's targets), which appears only very slightly larger.

Chapter 10.1 – July: Of Heroes and Crowns

July nights may be short, but the period of darkness includes the time when the Hercules region is near the zenith. This gives us an opportunity to appreciate some of the jewels the region has to offer with a minimum of image-extincting and -degrading atmosphere in the way.

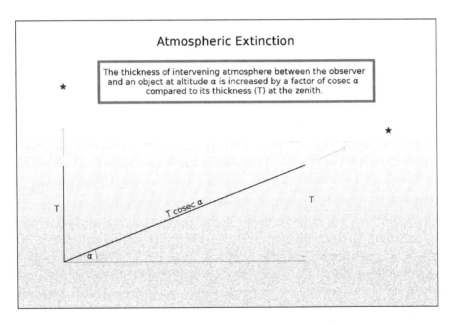

We'll start with M92, which is a lovely globular cluster in its own

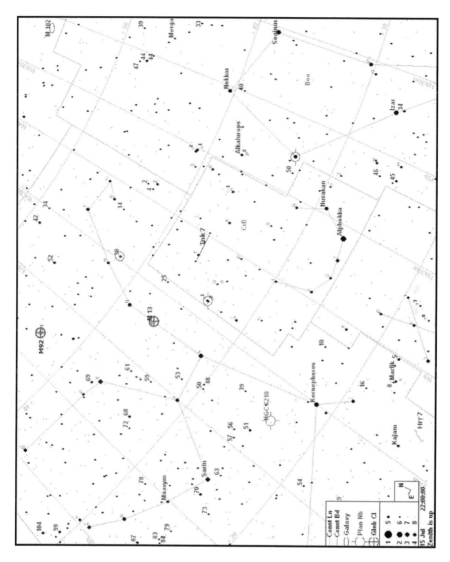

right, but which has a more famous one nearby, so tends to be ignored. Identify the keystone asterism in Hercules and start at the star at its NW corner (η Her). Next, identify the 4th mag ι Her, which is 12.5° to the NE (it looks distinctly blueish in binoculars). You'll find M92 one third (approx. 5°) of the way from ι Her back to η Her. M92 is an easy object, its 150,000 or so stars shining with a combined mag of 6.4 in apparent diameter about one third that of the Moon. You are seeing light that is the same age as the world's oldest ceramic, the Venus of Dolní Věstonice: it left there 26,700 years ago, about the time our ancestors discovered how to use plant fibres to make clothes and other artefacts!

M92's famous neighbour is M13, which many regard as the finest globular cluster in the northern hemisphere of the sky, which is why it is called the "Great Cluster". The Hercules clusters work in "thirds": this time start at η Her and go one third of the way down the W side of the keystone towards ζ Her. You might even be able to detect it with your naked eye, if you use averted vision under a dark transparent sky. It's so bright that you can even observe this cluster under urban skies, although obviously it will not look nearly as spectacular as it does under a good dark sky. In 10x50 binoculars, it brightens appreciably towards the centre, giving it the appearance of a tail-less comet. This explains why this is object No. 13 in Charles Messier's list of objects for comet-hunters to avoid. However, Messier did not discover it; its discoverer was the man whose eponymous comet Messier hoped to recover: Edmund Halley. Its light is only 22,200 years old, so it is a lot closer than M92. It has around 300,000 stars and a diameter of about 145 ly.

From M13 we move along to the distinctly orange star, 30 Her (also known as g Her) which you will find 1° to the WSW of σ Her. Over the last few years, the magnitude of this Semi-regular Variable (SRB) star has been varying between about 4.4 and 5.5 with a its period swinging between 73 and 93 days between maxima. 30 Her is about 230 times the diameter of the Sun, i.e. Earth would orbit inside it. Although it's so much bigger than the Sun, its mass is similar, and it is an example of how our star will evolve when it reaches its red giant stage. In the next stage of its evolution, 30 Her will lose its outer shell as a planetary nebula, and its core will decay as a white dwarf.

Even with your naked eye, you can see that ν CrB is a double star, so it is very easy to split in small binoculars, where it appears as a 6 arcmin wide, distinctly yellow pair, both of which shine at just under 5th mag 5. ν CrB is an optical double, which means that the stars are not gravitationally connected but are a chance line-of-sight pairing. The more northerly star, ν-1, is 555 ly away, which is 10 ly further than ν-2. By coincidence, they are very similar in nature; each of them is an advanced giant star of about 2.5 solar masses, but the brighter one is about 750 million years more advanced in its evolution, making it larger and intrinsically brighter.

Unlike ν CrB, our next object, δ Boo, is a true double star. The mag 3.5 primary is an orange-yellow star. The whiter 8th mag companion is just over 1.5 arcmin away. You are looking at a star that is similar to our Sun, so it is one that is studied by astrophysicists. Whereas 30 Her shows what our Sun will become, δ Boo's companion shows, from a distance of 117 ly, what it is now.

Our penultimate target is a fine group of stars, Tonkin 7. Go back to ν CrB, and pan 4° to the NW and find 5th mag τ CrB, the brightest star in a straight chain of five stars, about 2.5° long, running EW. The central one is the dimmest at under 7th mag (all the rest are brighter than 6th mag), but it also happens to be a very easy optical triple star. Note how the stars at each end of the chain are a deeper yellow than the others, and how the star next to the E end appears comparatively white.

An early advocate of binocular astronomy is the American author, Phil Harrington, who defined several asterisms, of which Harrington 7 is one of my favourites. Start off at Kajam (ω Her) and go 2° W to a golden yellow 8th mag star. Harrington calls the chain of 7th to 9th mag stars that zigzags 1.3° to the SE the "zigzag cluster. It's a fine example of pareidolia: Harrington sees a zigzag, but I see a long-tailed tadpole, and others have reported seeing things as diverse as a dragon and a flower. What image does it conjure up for you?

Chapter 10.2 – July: Pole Dancing

This region of the sky is, of course, visible all year round, but the objects here that we are looking at this month are as well placed after dark in July as they are at any other time of year. The word "pole", in this context, comes from the Greek πόλος (pólos), which means an axis of rotation (as opposed to the context of a long stick, which comes from the Latin palus, a stake).

The North Celestial Pole (NCP) is the point on the celestial sphere that lies directly above Earth's geographic North Pole. There are at least two other north poles of interest to astronomers. One is the Galactic North Pole, which does not interest us here; it lies in Coma Berenices, near the star 31 Com. The other is the Pole of the Ecliptic. Owing to a phenomenon called Precession (of the Equinoxes), Earth's rotational axis "wobbles", with a period of 25,772 years, and so the NCP sweeps out a circle on the celestial sphere. This circle has a radius of about 23.5° (the angle that Earth's axis is tilted with respect to the Ecliptic (the apparent path of the

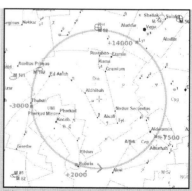

Fig 10.2.1 The changing position of the North Celestial Pole

Sun through the sky, so called because lunar and solar eclipses can only occur when the Moon is on that path). The centre of this circle is the North Ecliptic Pole. We shall meet it later. The consequence is that, at different times, different stars are nearest the NCP. When the Egyptians were building their pyramids, their pole star was Thuban (α Dra). In 1200 years, it will be Vega (α Lyr), and in just over 25000 years time it will be Polaris (α UMi) again.

Let's get our eyes in practice with a very pretty group of coloured stars, Tonkin 2, that span just over 1°. In the middle of the group is the blue-white 4th mag κ Dra, a very hot, very luminous star. This B-type star is more than 500 times as luminous as the Sun and has a surface temperature of 14,000 K (the Sun's is 5,778 K). The two orange-yellow stars to the N are K-type stars. The brighter of the two, 6 Dra, is dim by comparison, a mere 300 times as luminous as the Sun, and only 4,300K. 4 Dra, the star to the S of κ Dra is about the same luminosity as 6 Dra, but is even cooler (a mere 3940K), which is why it looks redder. 4 Dra also has a variable designation, CQ Dra. It is a long-period pulsating variable star with a small magnitude range (4.9 to 5.1) and an irregular period.

Next, we will visit two pairs of double stars. To find the first pair, OΣΣ123, start at Thuban and follow the chain of stars that winds 4° to the W. The components are close in brightness (mag 6.6 and 7.0) but well-separated (69 arcsec). This makes them very easy to split with small hand-held binoculars. The 'OΣΣ' refers to Otto Wilhelm von Struve's double-star catalogue. The Von Struves have been an eminent astronomical family for several generations, beginning in the 19th Century with Friedrich Georg Wilhelm von Struve (Otto Wilhelm was one of his eighteen children) and continuing to this day. Our next target is one of Friedrich's.

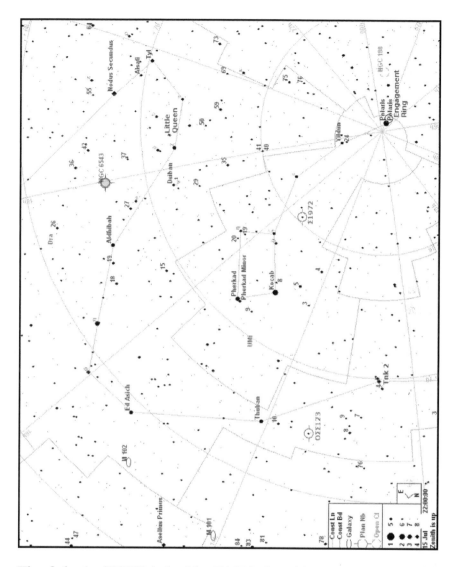

The father's (FGW's) double, Σ1972, is a bit more difficult. The brightnesses are similar (mags 6.6 and 7.2) but their separation is much smaller (31 arcsec). The easiest way to find it during summer evenings is to put both ζ UMi and ε UMi high in the field of view. You'll find Σ1972 slightly to the left of a point that is about 1.5° NW of the middle of the line joining ζ UMi and ε UMi. Look for a little equilateral triangle of 7th-ish mag stars. The one we want is the brightest, southerly, apex of it. It is a good test of the optical quality of both your eyes and your binoculars, so if you are having trouble splitting it, mount your binoculars and make sure they are perfectly focused.

Our first pole is near Polaris, which is more than merely a convenient marker for the North Celestial Pole (NCP). It is also part of the very pretty Engagement Ring asterism, which can also help us locate the NCP far more precisely. Your binoculars will show you an oblate circlet, with a long axis nearly 1° across, of 8th and 9th mag stars. Polaris blazes as the "solitaire diamond" in this slightly kinky ring. The kink exists because one of the 8th mag stars opposite Polaris in the circlet is slightly displaced away from Polaris, which bisects the line joining it and the NCP. This knowledge enables you to get a far more precise determination of the location of the NCP, which you may find useful, for example, if you are polar aligning an equatorial telescope mount.

Higher in the sky, you will find χ Dra and, in the same field of view, 1° towards Alsafi (σ Dra), you will find a little triangle of 7th mag stars. Either side of the star at the E apex (the one nearest τ Dra), you will see a pair of fainter 8th mag stars that create a trapezium of which the triangle is a part. These five stars form the characteristic "W"-shape of the five brightest stars of Cassiopeia, which is why this asterism has the common name "the Little Queen". Now look on the φ Dra side of the asterism, where there is a slightly fainter star that creates an almost-equilateral triangle with the two 8th mag ones. The Little Queen is part of a group that consists of a triangle within a trapezium within a triangle.

Last of all we come to the promised "other" pole: NGC 6543 marks the position of the North Ecliptic Pole, that point on the northern half of the celestial sphere that is the centre of the Sun's annual journey. Identify Nodus Secundus (δ Dra) and Aldibah (ζ

Dra); NGC 6543 is offset a degree towards Aldibah from the point mid-way between them. It looks like a faint star at first, but closer examination will reveal two important differences. First of all, it has a slight green tinge to it and secondly, unlike the stars around it, it "blinks" if you change between direct and averted vision. This is a planetary nebula, the glowing shell of gas that is the cast-off remnants of a dying star, and it is commonly called the Cat's Eye Nebula. Binoculars won't reveal any of the detail that imbues it with this name. It was discovered by William Herschel in 1786 and, nearly 100 years later, it was the first planetary nebula to be observed with a spectroscope. This revealed that its spectrum was not continuous like that of a star, but consisted of a few bright emission lines, showing that it consists of a rarefied ionised gas.

Chapter 11.1 – August: Swanning Around

Because it lies against the Milky Way, this region of sky has many open clusters and the sheer density of stars makes it inevitable that we should be able to find some asterisms that look like replicas of other objects. A problem with observing open clusters in the dense part of the Milky Way is that it is not always easy to distinguish them from the background stars, although this is generally easier at lower magnifications, so our binoculars give us an advantage for many of them.

1½° NW of ρ Cyg lies an unremarkable yellow star, 71 Cyg. This is the tip of an asterism, the Dart, that is our first "replica": it looks just like a smaller version of the brighter stars of Sagitta, the Arrow, with the two stars that represent the fletching slightly more than 1½° S of 71 Cyg.

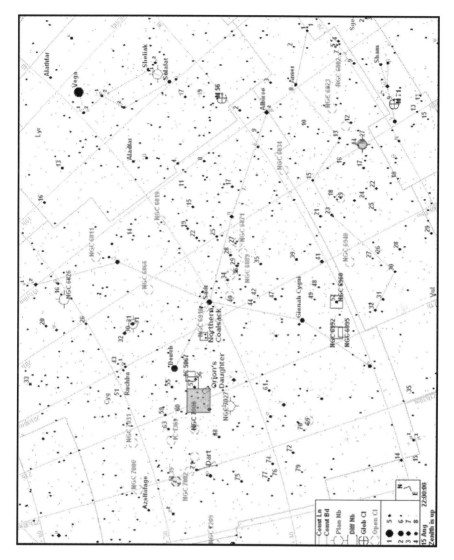

2° N of 71 Cyg is the sparse but bright (mag 4.6) cluster, M39, which you might even detect with your naked eye if the sky is dark enough. What you see with 10x50 binoculars will depend on your sky conditions. Under a bright suburban sky, you may only see a trapezoidal grouping of half a dozen stars, but try a rural sky and you should see at least three times as many and a change in shape from trapezoidal to triangular area as it grows to an area approximately the same angular size as the Moon.

Very slightly to the east of a line from Deneb (α Cyg) to Sadr (γ Cyg) is a dark patch of sky, B348, one of several regions that is sometimes called "the Northern Coalsack". The "B" is for the 19th Century American astronomer, E.E. Barnard, who we shall meet more fully in September, when we look at more of these dark nebulae. B348 shows you what a dark nebula looks like; pan around the Milky Way and you should find several more. Detecting them is a skill that will be useful later.

Our next target is 1° N of the middle of a line joining Deneb and δ Cyg. You are looking for a naked-eye double star; these are 31 Cyg (also designated o-1 and V695) and 30 Cyg. When you use binoculars on this double, you should see that the 4th mag southerly one, 31 Cyg, is itself a double star with an easily split, brilliant white, 7th mag companion 1.8 arcmin to the S. These three stars are not gravitationally related; what you are witnessing is an optical line-of-sight grouping.

The North America Nebula (NGC 7000) is difficult to see unless you have a dark transparent sky, when it can become visible to the naked eye, even with direct vision. What you'll see is a slightly brighter patch of sky, not the North America shape you see in images, unless you have a good dark site. Its centre is about 3½° ESE of Deneb. It responds very well to a UHC filter. If you have one, try holding it over one of the binocular's eyepieces. Another way I find useful (and the reason I tried to get you looking for a dark nebula earlier), is to try instead to detect the dark nebula that forms the "Gulf of Mexico". Once you have that patch of darkness, the brighter sky becomes more apparent. NGC 7000 is about 2° wide – which, at its distance of 2,200 ly translates to about 100 ly wide. As with so many nebulous objects, the characteristic shape is given to it by the intervening clouds of dust.

Our next "replica asterism" lies just on the SW edge of the North America Nebula. 2° SE of Deneb lies a 6th mag star, 56 Cyg. A little more than 1° SE of 56 Cyg, you'll see a pair of 7th mag stars about 10 arcmin apart. These are the equivalents of Saiph and Rigel in a tiny Orion-shaped asterism of seven stars, which I know (with apologies to that wonderful film director, Sir David Lean) as Orion's Daughter. About half a degree to the N you will see the Betelgeuse and Bellatrix equivalents, slightly fainter than our first two stars and slightly wider spaced. In between these two pairs of stars is the line of three 9th mag stars that makes the "belt". The left-hand shoulder (pseudo-Betelgeuse) marks the tip of "Florida" in the North America Nebula, and you can use this as a guide to the location of the dark "Gulf of Mexico" patch (above).

One of the most attractive double stars in the entire sky is found in the head of Cygnus, the Swan, yet the meaning and origin of its name are unknown: Albireo (β Cyg). The two components have a separation of 34 arcsec, making them a good test for 10x magnification binoculars, so if you don't initially see two stars, make sure they are perfectly focused and keep them as steady as possible. Its beauty lies in contrast between the gold of the 3rd mag primary and the sapphire blue of its 5th mag companion. For a long time, Albireo was suspected to be to be a true binary star (and many sources still assert that it is) but, in 2018, data from the European Space Agency's Gaia spacecraft are definitive in identifying it as an optical (line-of-sight) pairing only.

The Dumbbell Nebula (M27) is by far the easiest planetary nebula for small binoculars. Identify γ Sge and pan a bit more than 3° towards 15 Vul where, even in moderately light-polluted suburban skies, you will find the ghostly glow of a tiny rectangular cloud. This is the Dumbbell Nebula (aka the Apple Core and the Diablolo). Although it initially appears rectangular, give it a bit of time, allow your eyes to relax, and use averted vision and you should detect the slight "waist" in the middle that gives it its common name. Planetary nebulae, which form from the death of Sun-like stars, are typically 1 ly in radius; the 1,360 ly distant M27 is no exception.

Chapter 11.2 – August: Nothing But Stars

We're staying in the same region of sky for the second part of this month's exploration. There is so much here worth looking at, and the Summer Triangle region is very well placed on late summer evenings (hence its name, of course). There's nothing here that you cannot see with 50mm binoculars; my decision on what to include in this section is merely my opinion on which objects benefit most from greater aperture and magnification. See if you agree with me.

If you put Sadr (γ Cyg) at the N of your field of view, you will find a 7th mag cluster near the centre, just less than 2° S of the star. Although M29 ("the Cooling Tower") is a fairly unremarkable object in smaller binoculars, on a good night your 15x70s will resolve up to a dozen stars against a diffuse background that is similar in appearance to a globular cluster under high magnification. Some of these brighter stars appear to make the shape of a power station cooling tower, hence one of its common names. Like many open clusters, its distance is not precisely known, but is in the range 4,000 to 7,200 ly. It is believed to be about 10 million years old, and its brightest stars are hot blue giants, each with a luminosity of over 150,000 Suns. Seen from there, the Sun would be invisible, at mag 16, in our binoculars!

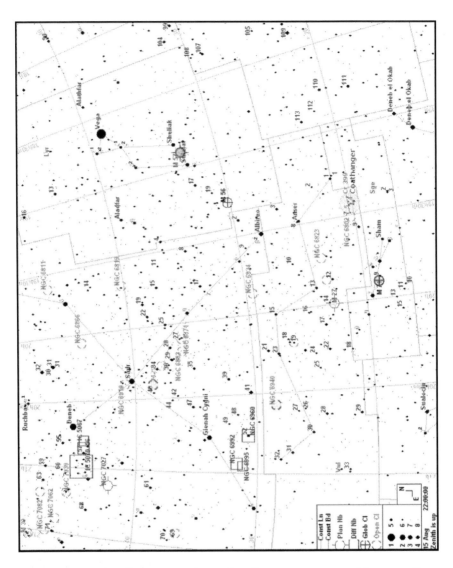

61 Cyg was the first star to have its distance measured by parallax. In 1838, F.W. Bessel measured its parallax (i.e. half its apparent angular shift in position over 6 months) as 260.5 milliarcseconds (mas) – 72 millionths of a degree – giving a distance of just over 10 ly, very close to our modern measurement of 11.4 ly. This was the first reliable distance measurement to any star other than our Sun. Imagine that Deneb (α Cyg), Sadr and Gienah (ε Cyg) are the apexes of a parallelogram; 61 Cyg is located at the 4th apex. It has

a large proper motion (motion relative to the celestial sphere) of 5.22 arcsec/year, hence one of its common names, "Piazzi's Flying Star". Your binoculars should reveal it to be a pair of orange stars, 31 arcsec apart, with magnitudes of 5.2 and 6.0; 61 Cyg is a true binary star, not a chance line-of-sight pairing.

NGC 6940 is a large bright open cluster that rewards patient observation. To find it, locate 52 Cyg and pan 3° SW, where you will find a soft elliptical glow ½° across. In 70mm binoculars from a dark site, you should be able to resolve ten or more stars against a mottled glow. If you have larger binoculars, do try them: they should reveal several strings of stars within the cluster. The light from this 800-million-year-old cluster left it in about 600BCE, when the Hanging Gardens of Babylon, one of the Seven Wonders of the Ancient World, was being built.

A mystery object: is it a tight open cluster or a loose globular cluster? After many years of dispute, the consensus is now that M71 is actually a globular. You can find this puzzling object by centring your binoculars on the mid-point of a line joining δ and γ Sge. M71 will be in the field of view, about ¼° S of centre. It's quite faint: in 15x70 binoculars with direct vision it looks like a badly defocused star, and if you switch to averted vision, you can confirm that it's an extended object because it seems to grow a bit. M71 is thought to be only about 9 billion years old, which is relatively young for a globular cluster which, theory dictates, formed around the same time as our galaxy.

The Coathanger is a popular star-party piece, because it looks like an everyday object and because at 1.5° across, it will stay in the field of view of mounted binoculars long enough for several star-party observers to see it. It is properly designated as Collinder 399, but is also known as Brocchi's Cluster and Al-Sufi's Cluster.

You'll find it 5° S of Anser (α Vul) and 4° NW of Sham (α Sge). Even 20 mm toy binoculars will reveal the ten brightest stars that give this asterism (the Hipparcos satellite data show that it's not a true cluster, just a random line-of-sight grouping of stars) its common name. It was first recorded in Persia, in the 10th Century CE, by Abd al-Rahman al-Sufi, but has probably been known for much longer than that. One of the alternative common names, Brocchi's cluster, arises from its use by an American amateur astronomer, Dalmiro Brocchi, for calibrating photometers. It is useful for this because stars with a wide range of magnitudes fit into a single field of view. You could find the limiting magnitude of your binoculars using the chart in Appendix A.

We'll finish with a planetary nebula, the Ring Nebula. Object number 57 in Charles Messier's catalogue of non-comets, it lies half way between Sheliak (β Lyr) and Sulafat (γ Lyr), which makes it easy to locate, but it can be tricky to identify in 15x binoculars. You're looking for what appears to be an out-of-focus 9th mag star; so don't expect to see it as a ring with dark space in the middle unless you have exceptional sky conditions.

Chapter 12.1 – September: Something Fishy

The environs of Pisces are often considered to be a bit of a binocular void, owing to the paucity of bright deep sky objects (it boasts exactly one Messier object: the galaxy, M74, which is remarkably difficult in 50mm binoculars), and it doesn't even have any stars brighter than 4th mag. Yet a little perseverance is rewarding, as it is in almost any part of the sky. Get ready to enjoy some coloured stars and to hone your double-star observing skills.

We'll start with the Pisces Parallelogram (Tonkin 3), an asterism that needs binoculars to reveal its colours. It is a 3° x 1° parallelogram of 4th and 5th mag stars that has 27, 29, 30 and 33 Psc at its corners. The deep orange star at the SW corner is 30 Psc and, diagonally opposite, there is a star at the opposite end of its evolution, the sapphire blue 29 Psc. The other two corners of the parallelogram are a comparatively insipid yellowish colour in binoculars. In 10x50 binoculars the northern part of parallelogram appears empty of stars, but if you have very dark skies or access to a larger aperture instrument, you'll see that this is an illusion as a few fainter stars come into visibility.

Approximately 10° NW of Tonkin 3, we find the southern circlet of Pisces. On the eastern side of this circlet you will see a deep red carbon star, TX Psc (aka 19 Psc). The magnitude of this ruby

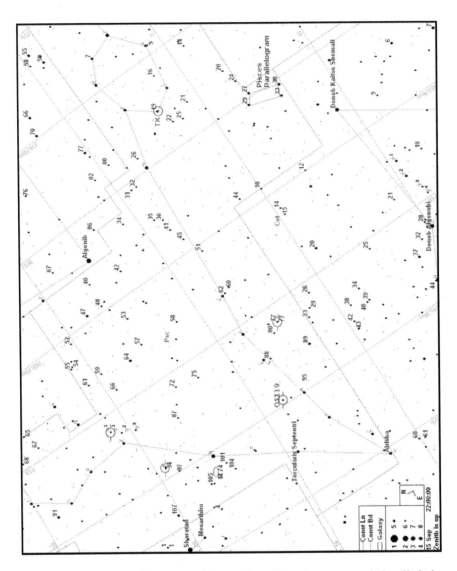

droplet is usually given as 5.0 but, like all carbon stars, it is slightly variable (4.8 to 5.2). In TX Psc, the variability is due to a cycle of expansion, during which its luminosity increases, then contracting and throwing off some of its sooty carbon atmosphere. This obscures the star and decreases its luminosity. This cycle has a very long and irregular period.

You may also notice a different, and entirely unrelated, kind of variability with this red star: if you stare at it, it looks brighter than if you merely glance at it. This is called the Purkinje effect, and you can experience it to some extent in all red stars if you have sufficient aperture. It is something to be aware of if you are visually estimating the magnitude of red stars because, unless you only take quick glances at them, you will systematically overestimate their brightness.

We'll now move into the northern branch of the Pisces asterism to find ψ-1 Psc degree and a half from χ Psc towards η And. The two brilliant white stars (mags 5.3 and 5.6) that comprise this double star are separated by 30 arcsec. This is quite close for a hand-held 10x50 binocular, so you will need to use all the "tricks" if you are going to split it. First of all, make sure that both sides of the binocular are in perfect focus. Secondly, mount the binocular; if you aren't able to do this, try some of the suggestions in Chapter 3. Thirdly, keep the target in the middle of the field of view: this is the region that is least affected by optical aberrations. Lastly, if you usually wear spectacles, put them on; they will probably help.

We go back S for 77 Psc which is another close double star, this time with yellow-white components at mags 6.4 and 7.2 respectively and separated by a marginally easier 33 arcsec. To find it, start at ε Psc and hop 3° S via the slightly brighter orange-yellow 73 Psc.

Our next target is another of Otto von Struve's (see July (2)) doubles. Hop 2° N from μ Psc to find OΣΣ 19, an easy, contrasting double star. The primary is a 6th mag orange star and its companion, a very easy 69 arcsec away, is an 8th mag yellow-white star.

Our last double star is the easiest of them all. ρ and 94 Psc shine at mags 5.4 and 5.5 respectively and, if the sky is dark enough for you to see them with your naked eye, even that will show you that there are two stars there: they are a whopping 7.5 arcmin apart. ρ is a yellow-white star, but 94 has a lovely orange light, and binoculars enhance this difference.

Chapter 12.2 – September: Embracing the Dark Side

From Cygnus down, through Sagittarius to Centaurus, the Milky Way appears to be split into two by what has become known as the Great Rift. These darker regions do not seem to have featured much in European astro-mythology, but things are very different in the southern hemisphere. In both Australasia and South America, these regions held significance for the native peoples. For the Incas, these dark nebulae formed various animals, whilst for Australian Aborigines, a single Great Emu stretches from Scorpius to the Coalsack Dark Nebula (which forms its head) in Crux.

Many of these dark nebulae have Barnard (B) numbers. The B is for E.E. Barnard, nicknamed "The Man who Never Slept" by his colleagues, who catalogued many of what had been thought since William Herschel's time, to be "holes" and "lanes" cutting through the Milky Way. By carefully studying photographs of these that he took with the 40" refractor at the recently closed (October 2018) Yerkes Observatory in Wisconsin, USA, he realised their true nature: clouds of gas and dust that obscure our view of the Milky Way.

Observing them can be a bit of a knack – you are trying to train your eye to look for the absence of something. They are badly

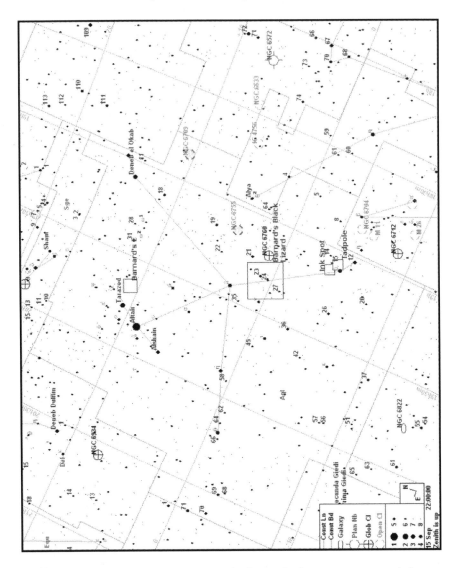

affected by light pollution, so wait for a dark, transparent night, at least until you have more experience and know what you're looking for.

The distinctive pair of dark nebulae that comprise Barnard's E (B142 & B143) is easy to find as it lies just 1° W of Tarazed (γ Aql). The nebulae lie in a rich starfield, which makes them easy to identify; they form either an uppercase letter E or an underlined C, whichever suits your fancy.

You will need a really transparent dark sky for Barnard's Black Lizard (B138). Begin at δ Aql and pan 2.5° SW to the 5th mag 23 Aql. Give your eyes time to take in the detail of what is in the field of view, and you should notice that 23 Aql is situated in a narrow, almost starless, band that runs NE to SW, before it sweeps eastward, passing 1° W of 27 Aql. This narrow band is the Black Lizard. Notice how its E edge is quite well-defined but, on the W side, it tends to blend into the Milky Way background.

Use λ Aql to find 15 Aql. You'll find the Ink Spot (B135) ⅔° E of 15 Aql. It's quite large – about 13 arcmin across – and I see it more as a mushroom than an ink spot, as it has a thick stalk extending to the SW. Immediately to the east of it is a smaller "ink smudge" that is not quite as dark. That is B136.

Go back to λ Aql and look at the spot 2/3° NW of it (½° SSW of 15 Aql). There you'll see a very dark short (¼°) streak running NE-SW. This is the Tadpole (B132), the last of our dark nebulae in this set.

For a bit of "light" relief, we'll finish off with a couple of open clusters. Identifying open clusters against the background Milky Way can be tricky because they tend to merge into the background. The trick is to use just the right amount of magnification; too much and the cluster no longer looks like a distinct cluster. These ones are just about right for 15x70 or 16x70 binoculars as long as you have a dark sky. Our first one, NGC 6755, lies 3° E of Alya (θ Ser). It's quite large (about 15 arcmin across), but you may not be able to resolve any stars; they just remain as a ghostly glow against the Milky Way. The next, NGC 6709, is slightly smaller but a little brighter. It lies 5° SW of Deneb el Okab Australis (ζ Aql). You definitely won't be able to resolve any stars, but you should see another spectral glow against the Milky Way stars.

Chapter 13.1 – October: A Princesses and a Hero

A huge expanse of the northern sky is devoted to a single Greek myth, that of Perseus and Andromeda. Here we have not only the hero and the princess whom he rescued, but also flying horse, Pegasus, that Perseus used in his task, the terrible sea monster that would have devoured Andromeda, and even the evil petrifying eye of the Gorgon. Looking down on the action are Andromeda's parents, King Cepheus and Queen Cassiopeia.

This group teaches us how the ancients made "sense" of the sky, and can help us to do the same. Objectively, what do you see? Several thousand points of light. If you join some of the brighter points into stick figures, you may at least remember the spatial relationships of the group – constellation – you have just made. Then, linking these groups, you weave a story. The story binds the constellations together in your memory and imagination so that, when they reappear after having been absent from the night-time sky for a month or two, the relationship is as strong as ever. When social knowledge, which this is, is primarily preserved and communicated orally, stories are the ideal vehicle of communication.

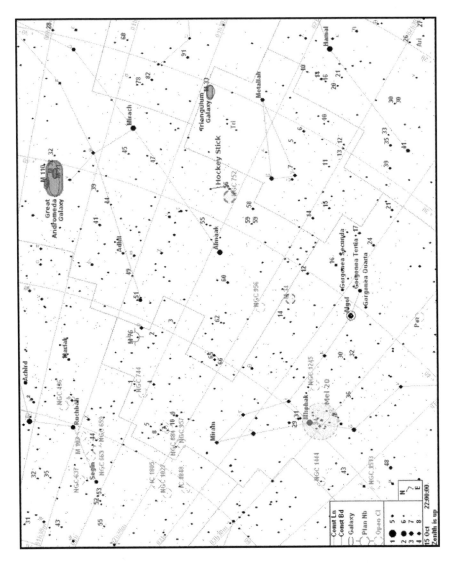

We'll begin with a stunningly beautiful target for binoculars. Melotte 20, otherwise known as the Alpha Persei Moving Cluster, extends for more than 3° SE from Mirphak (α Per). The stars here shine with an intense blue-rich whiteness that is indicative of their relative youth, probably about 60 million years. These hot young stars are mostly of spectral types O and B, and this accounts for the binocular view of them, in which they have been said to "sparkle like diamonds on black velvet". It's called a "moving cluster"

because all the stars are moving the same way relative to us, changing their position by 33 milliarcsec per year.

Algol (β Per) is a type of variable star called an "eclipsing binary". There are two stars orbiting their common centre of mass. The solar system is close to the plane of their orbit, and the stars are not the same brightness, so their combined brightness falls when the fainter star passes between us and the brighter one. This happens every 2.85 days, when its magnitude drops from 2.1 to 3.4 for a period of 10 hours. In our Perseus and Andromeda legend, this is Medusa's petrifying eye. Its name comes from the Arabic "Ras al Ghul" (yes, there is a Marvel Comics villain of the same name!), which means the "Demon's Head". If you are starting out in variable star observation, Algol is a good choice to cut your teeth on because of its short period and the fact that it is circumpolar from the UK. This month is a good time to begin as it is high in the evening sky, and will remain so for several months.

5° from Algol in the direction of Almaak (γ And) you will find a fuzzy patch about ½° across (same size as the Moon appears); you may even be able to see it with your naked eye if you use averted vision on a transparent night. This is M34, which is a lovely sight in any size of binocular. Your 10x50s can, depending on your sky conditions, show a dozen or so stars, the brightest of which form a "bent 'H'" shape. At a distance of 1,400 ly, M34's ½° diameter translates to about 14ly. It's about 220 million years old.

We'll continue with another lovely open cluster. Find β Tri (which is actually the brightest star in Triangulum) and place it at the SE of the field of view. NGC 752 (Caldwell 28), should be visible NW of centre, just N of a close pair of 6th mag deep yellow stars (56 And). NGC 752 is about twice as wide as M34, which makes it an easy object even in small compact binoculars like an 8x20.

However, it really benefits from increased aperture and magnification so, if you have larger binoculars, give them a try. With a pair of 15x70s, you should be able to resolve at least twenty stars against a background glow that is nearly twice the apparent size of the Moon, and 25x100s should reveal over 30 stars. It's only 1,200 ly away, making NGC 752 a relatively close cluster, which explains why so many stars are easily resolved. It's also very old for an open cluster: about 2 billion years. This explains the unusually large number of yellow stars; open cluster typically contain hot blue stars.

Return your attention to 56 And. You will see a straightish chain of stars of 6th and 7th mag stretching for 1.6° to the NW. They form an asterism that is shaped like an (ice-)Hockey Stick, with 56 And as the blade and NGC 752 as an unfeasibly large puck. This is a line-of sight association: NGC 752 is four times further away than 56 And.

Next, place the distinctive mag 2.0 yellow star Mirach (β And) at the SE of the field of view and find μ And near the other side. Place μ And where Mirach was, and the elliptical shape of the Great Andromeda Galaxy (M31) should now be where μ And was. The brighter glow of the nucleus can cut through quite severe light pollution but, under good suburban skies, you should be able to see the extent of the spiral arms and make out the two companion galaxies, M32 and M110, which appear as out-of-focus stars. In suburban skies, you may even be able to see M31 with your naked eye: for most people, it is the most distant thing they can see without optical aid. The light you are seeing is over 2 million years old; this is ten times as old as our species and it left the galaxy before our Australopithecine ancestors became extinct!

We have left the Triangulum Galaxy (M33) until last because you will need dark-adapted eyes, as well as a transparent sky, if you are to find it. Return to Mirach and scan the region that is 7° of the way to Hamal (α Ari). Here you will find the large, faint, circular ethereal glow of M33. It is face-on to us and spread out over an area of sky that is even larger than NGC 752, making it very faint. Although it is listed as being of mag 5.5, this is its integrated magnitude, i.e. the magnitude it would appear if the light was concentrated into a star-like point. Because its light is so spread out, it is a low surface-brightness object, which means that there is very little contrast between it and the background sky. This low contrast, combined with its size, makes it one of several objects that is easier to see in a pair of 10x50s than in a small, equivalent-priced, telescope.

Chapter 13.2 – October: Globes and Corpses

In this set of objects, we look at extremes of age in our galaxy. Globular clusters are believed to have formed when our galaxy did (and some may even be the cores of small satellite galaxies that have had their surrounding stars ripped off by the gravity of the Milky Way). As such, they contain what must be amongst the oldest stars in our galactic environment.

At the other end of the scale, planetary nebulae form at the end of a star's life and last for only a few tens of thousands of years, a mere blink of an eye in cosmic terms. They have nothing to do with planets, but are the ejected gas from a dying star that glows because it has been energised by ultraviolet radiation from the hot luminous core of the stellar corpse. They get their name from their appearance in early telescopes: they were cloudy patches (nebulae) that had a disc-shape like a planet.

Our first object in this set is the globular cluster M2, which is a doddle to find: it is N of Sadalsuud (β Aqr) and W of Sadalmelik (α Aqr). It stands out in an otherwise sparse region of sky, and you should be able to see this "nebula without stars" as Charles Messier described it in the 18th Century. If you use averted vision (by directing your gaze away from it), it will appear slightly larger and may adopt a very slightly oval shape. In exceptionally good

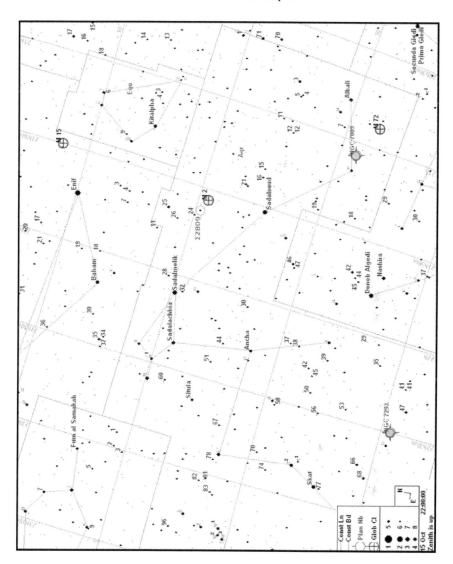

sky conditions, you may notice that it has a slightly granular appearance.

Our next target is another globular cluster, M15, which is also easy to find. Try extending a line from Baham (θ Peg) to Enif (ε Peg) another 4° to the NW. It will look slightly larger and less oval than M2. Images reveal it to be 18 arcmin wide but don't expect see it as more than half this width in 70mm binoculars. Most of its stars are in a dense core that not even the Hubble Space Telescope can fully resolve, and the outer regions consist of sparse faint stars beyond the range of 70mm binoculars.

Go back to M2, and from there go slightly more than a degree to the ENE to a 6th mag star. This is the primary component of a double-star system, Σ2809. The fainter (9th mag) companion will be a stiff test of your observing technique, as it is only 31 arcsec SSE of the primary. To have a good chance of success, wait until you have a dark, transparent sky, and mount your binoculars and then focus them as critically as you can. If there is any turbulence in the atmosphere, the secondary star may not be permanently visible, but may seem to be there one moment and not the next.

You're going to need a good southern horizon, a transparent sky, and full dark-adaptation for the last three objects. Pan 10° S from σ Aqr, where you will find υ Aqr (if you cannot see either of these stars with your naked eye, you may struggle with these final objects). A little more than 1° to the W, you will find a well-demarcated faint circular patch that appears a little smaller than the Moon; this is the Helix Nebula, NGC 7293. You won't get even a glimpse of the helical structure that gives it its common name, but you may just notice that its periphery is brighter than the centre. Averted vision may hint, unconvincingly, at some detail. This is most noticeable at the N of the nebula. At a distance of 450 ly, it is the nearest planetary nebula to Earth, which accounts for its apparent size.

Our penultimate object, the Saturn Nebula (NGC 7009) demands the same observing conditions as the Helix did, being so near the horizon murk in British skies, but it has the advantage of being very easy to locate. Start by finding v Aqr, then go 1.3° W to a small nebulous patch that looks like a defocussed star; you may notice a hint of a greenish tinge to it. It was discovered by William Herschel in 1782, and is the object to which he first applied the term "planetary nebula" on account of its appearance in his telescope as a ghostly, nebulous planet. He also thought that the nebulous substance surrounding the progenitor star core might be surrounding gas that was condensing into planets.

Our final target, the globular cluster M72 which, at mag 9.4, is the faintest of the Messier globulars, is never ideally placed from Britain: atmospheric extinction takes nearly half a magnitude off it, even in the south. Its location is easy, though. About 3° SSE of Albali (ε Aqr) lies a pair of 6th mag stars. One is directly N, the other directly W of M72, which looks like a slightly defocused star. If you ever attempt a Messier Marathon, you'll find that M72 is one of the more difficult objects, so it's useful to be able to locate and identify it easily in less than ideal sky conditions.

Chapter 14.1 – November: Pareidolia

Asterisms are those chance line-of-sight associations of stars that evoke mental images of a familiar item, a phenomenon called pareidolia. It should not come as a surprise that these seem to be most abundant where the stars are most abundant: in or near the Milky Way. We have already met some of these northern circumpolar ones in May; there are another four here.

We'll begin with an asterism that is named for the prolific Canadian binocular observer, the Franciscan friar, Fr. Lucien Kemble (1922-99). Take a line from Segin (ε Cas) through ι Cas leads to γ Cam, 8° further NE. Approx. 1.5° W of γ Cam, you'll find a yellowy-orange star, the semi-regular variable (mag 6.3 to 6.5) V805 Cas. V805 is the brightest of a 1.5° long group of ten 8th and 9th mag stars that form a diamond kite with a tail, which flows S towards Perseus. This is Kemble's Kite. Once you have identified it, take a look at the star at the northern tip of the kite: it is a double star that is very easy to split: both stars are mag 8.4 and are 103 arcsec apart. Can you detect any difference in their colours?

Camelopardalis doesn't have any stars brighter than 4th mag, which makes it rather indistinct in typical UK skies, so we'll again use the brighter stars of Cassiopeia to locate Kemble's next object.

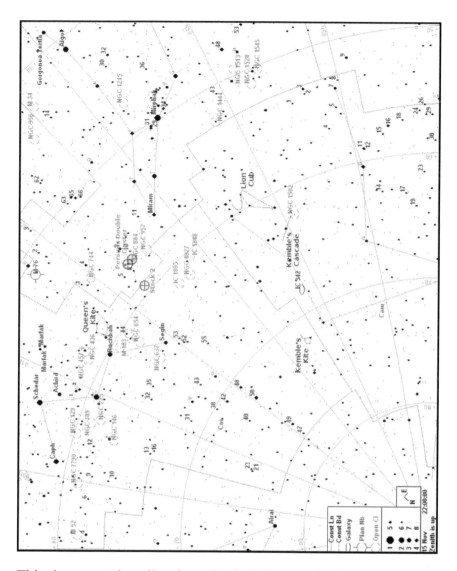

This time, we take a line from Caph (β Cas) to Segin, and extend it for the same distance into Camelopardalis. Kemble's Cascade is a nearly straight line of fifteen 8th mag stars with a brighter (5th mag) one in the middle. It extends 2.5° NW-SE and has an open cluster, NGC 1502 near the SE end. At this time of year, if you want to get the full effect, observe it as early in the evening as possible, then you can see it as a vertical ribbon waterfall with NGC 1502 as the splash-pool at the bottom.

Our next object is yet another asterism, the Lion Cub (Tonkin 5). It looks uncannily like the brighter stars of the constellation Leo. We need to find CS Cam and, unsurprisingly, we use Ruchbah as our starting point once again: CS Cam is 15° to the E. Another 1° E brings you to a near-perfect semi-circle of 8th-ish mag stars. Now, if you imagine that the semi-circle is a small version of the sickle asterism of Leo, then use chart below to identify the rest of it.

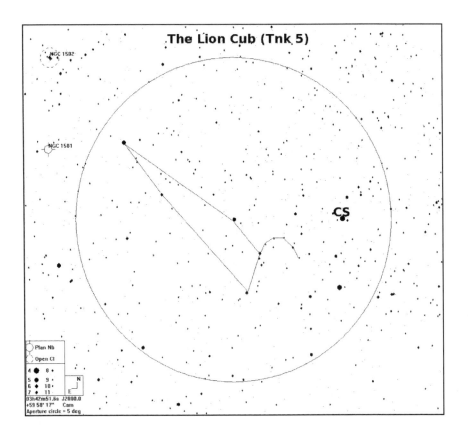

The Perseus Double Cluster is the only non-asterism in this set of targets. Once again, we'll use the stars of Cassiopeia to guide us. This time we take a line from Navi (γ Cas) to Ruchbah (δ Cas) and extend it for an additional double the distance between the two stars. In the field of view, you will see a very close pair of open clusters, NGC 869 and NGC 884. If you have clear dark skies, you might even be able to see them with your naked eye as a single fuzzy blob. The two distinct condensations of stars are easy to distinguish in small binoculars, but if you have larger ones, you can use them to reveal the varying colours and brightnesses of the stars that give the clusters a 3D appearance. Those stars are intrinsically very bright: if we were located there and tried to look back to our Sun, it would be much too faint to be seen in our binoculars!

From the more westerly cluster (the one nearest Cas), you will see a 2° curved chain of 8th mag stars leading to the N. These are your guide to Stock 2, the Muscleman Cluster. This large, faint cluster gets its common name from the asterism made by its brighter stars. These form a stick-man, with his feet to the E, standing in a bicep-flexing body-builder pose, and wielding the Double Cluster on a chain (the 2° chain that led us here). The connection between the Muscleman and the Double cluster is purely illusory: the Muscleman is only 1,050 ly away and the Double Cluster is nearly seven times as far.

We finish this month as we began, with a kite asterism. Yet again, we return to Ruchbah, but this time, locate yellow χ Cas, 1.5° to its SE. Another ¾° further on, there is a 7th mag star. This star lies at the middle of a slightly irregular pentagon of stars that has χ Cas as its brightest apex. This is the Queen's Kite. Its tail is the 1.6°-long chain of stars that leads S from the most easterly star of the pentagon. It's the varied colours of the stars that make this asterism particularly attractive. You should note that most of the fainter stars are white, but the two brightest stars are yellow.

Chapter 14.2 – November: An Autumn Jewel Box

There is so much to see in this part of the sky, that we return to it for the second set of November objects. Cassiopeia is very high in the sky during the evenings at this time of year, so atmospheric extinction is minimal, but the trade-off is that high-altitude observing can be uncomfortable if you are using "straight-through" binoculars. If you are going to make a habit of observing high altitude targets, a recliner is essential, a monopod and trigger-grip head helps, but a parallelogram mount makes high elevation observing a sheer pleasure.

We'll begin with a trio of open clusters.

To find Melotte 15, imagine that Segin (ε Cas) and ι Cas are two points of an equilateral triangle. Mel 15 is at the third apex, to the south of the two stars. You should see it as a large (20 arcmin), sparse open cluster with eight or ten of the brighter stars resolved against the background glow from the fainter stars of the cluster. You should notice that these brighter stars form a chevron near the centre of the cluster. If you have exceptional skies and access to a UHC filter, try putting the filter over one of the eyepieces. This might enable you to see a glimmer of the nebulosity (IC1805, The Heart Nebula) that surrounds and gave birth to the cluster. Its shape, which you can see in images, is sculpted by the stellar

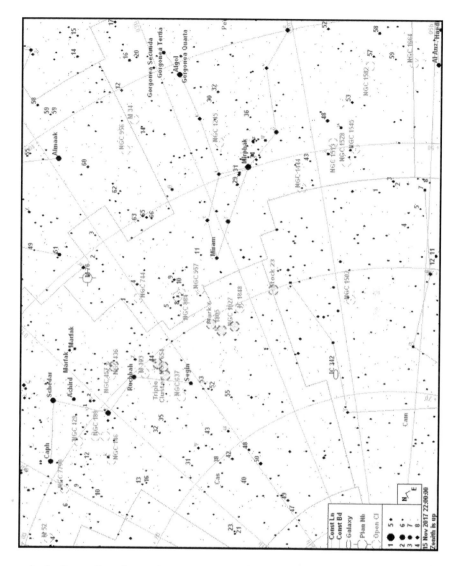

winds from the cluster's very young hot supergiant stars.

Our next cluster, Markarian 6, is a little less than a degree to the SSW of Mel 15. About ¾° S of Mel 15 you will see a solitary 8th mag star; Mark 6 is 11 arcmin W of that star. In 15x70 binoculars, you'll see a south-pointing arrowhead of half a dozen 9th mag stars. It's smaller and fainter than Mel 15, so you might think that it is more distant but, at 1,600 ly, it is actually not even a quarter the distance away.

The third cluster of our first trio is a lot more obvious than Mark 6. To find it, look slightly more than a degree to the E of Mel 15, where you will find a 7th mag star. NGC 1027, which surrounds that star, is nearly as big and nearly as bright as Mel 15, but it looks completely different. It looks very much like a comet with a halo (the glow from the unresolved stars of the cluster) surrounding a bright condensed core (our 7th mag star). It reinforces the illusion of distance that started with Mel 15 and Mark 6. It lies 4,000 ly away, intermediate between the other two clusters, but it is closer to us than the brighter Mel 15, and further away than the dimmer Mark 6.

A little more than 1½° W of CS Cam, in the direction of NGC 1027, lies a little trapezium of 7th and 8th mag stars. This is Stock 23, also known as Pazmino's cluster, named for the American amateur astronomer, John Pazmino, who popularised it in the 1970s. A little patience with 70mm binoculars will show you that there is much more here than the trapezium and you should be able to tease about a dozen stars out of the 10 arcmin diameter, ethereal glow that emanates from the unresolved fainter stars of the cluster.

Use Ruchbah (δ Cas) as your starting point, and pan 2° towards Marfak-East (θ Cas). Here you will find an easy, bright wide double star (5th and 7th mag, 134 arcsec apart). The brighter of the pair is identified on the chart as φ Cas. These are the eyes of the Owl Cluster (NGC 457), also sometimes known as the ET Cluster and the Dragonfly cluster. φ Cas is actually a foreground object and not part of the cluster. The Owl's (or Dragonfly's) body and wings (or, if you prefer, ET's body and arms) are comprised of 9th and 10th mag stars that cover a bit more than a ¼ ° spread, on the γ Cas side of the eyes.

We'll finish as we began, with a trio of open clusters, this time it's the Cassiopeia Triple Cluster. They're very easy to locate: just look 1° E of the midway point between Segin and Ruchbah, where you will see the most obvious of these clusters, NGC 663. You should notice a dark lane in the middle that separates the four brightest stars into two pairs. The brightest of the trio is NGC 654, which is between NGC 663 and the imaginary line between Segin and Ruchbah. Although it is brighter, depending on your sky conditions, you may find it less obvious than NGC 663 because it is so much smaller. The faintest of the trio, and the one that is most likely to evade detection, is NGC 659. You will find it next to the 6th mag star 44 Cas, in the direction of NGC 663. It's about the same apparent size as NGC 654, but is less than half as bright; with anything other than an ideal sky behind it, you'll probably need averted vision.

Chapter 15.1 – December: Diamonds and their Dust

The spectacular Pleiades, also known as the Seven Sisters and M45, are an easy naked-eye object, but put it in the field of small binoculars and it is as if a handful of diamonds had been tipped onto blue-black velvet. Under suburban skies, you should see at least forty of these hot, brilliant blue stars, and in dark skies it is easy to lose count of them. Look for the many subtle curves and chains of stars, especially Ally's Braid, a chain of 7th and 8th mag stars extending for nearly a degree S from Alcyone ("Ally"), the brightest star in the cluster.

The first artistic depiction of the Pleiades (M45) is in the 17,000-year-old Lascaux cave paintings. M45 lies about 450 ly away and imagery shows that they are surrounded by gas that reflects their starlight. This is not the nebulosity out of which they formed some 80 million years ago, but a cloud of gas they have encountered in their passage through space, and which they are destroying with their radiation.

Fig 15.1.1 Lascaux cave painting

The Lascaux paintings also depict the Hyades cluster (Melotte 25, Caldwell 41), which is next to

118

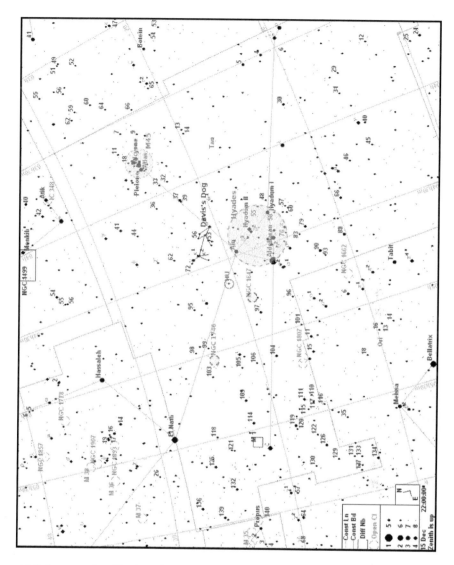

Aldebaran (α Tau), the reddish eye of the bull, which is a foreground star.

The Hyades will extend beyond the field of view of all but wide-angle binoculars. Expect to see at least thirty stars in this, the nearest (153 ly) open cluster to Earth. In Greek mythology, the Hyades were the sisters of Hyas. They were the children of the Titan, Atlas. After Hyas was killed by a lion, his sisters' tears became associated with the rains that coincide with the heliacal setting of the Hyades cluster in in spring: these stars are the "April rainers" in the song "Green Grow the Rushes, O".

Davis's Dog is the most dog-like of doggy asterisms. You'll find it 3° from the "northern eye" of the bull, Oculus Boreus (ε Tau), in the direction of the Pleiades. The star directly between Oculus Boreus and the Pleiades is ω-2 Tau, the nose of Davis's Dog, which extends about 3.5° towards El Nath (β Tau). This little asterism has fifteen stars of 8th mag and brighter, the brightest of which are the 4th mag υ Tau and the wide double star κ-1 and κ-2 Tau. These are part of the curved string of four that make its tail. A right triangle of three 5th mag stars form its head, and fainter stars form its stubby legs. John Davis, who discovered this asterism, saw a beagle, but I see it as something with much shorter legs and longer body, perhaps a dachshund. Which breed of dog do you see?

If you look 3° from of Oculus Boreus in the direction of El Nath, you'll find a (usually) 6th mag star, HU Tau. I say "usually", because, as HU indicates, this is a variable star, in this instance an eclipsing variable. The brightness of eclipsing drops the most during its primary eclipse, which is when the fainter companion star passes in front of the brighter primary star. HU Tau's primary eclipses last about 7 hours, during which its brightness drops by nearly a magnitude from 5.9 to 6.7. This happens every 2 days 1hr and 21 mins so, if eclipses are occurring in daylight when you read this, you will have 3 weeks at the most to wait before they are fully observable night-time events.

If you put Aldebaran at the SW edge of the field of view, somewhere (depending on the field of view) on the opposite side you will see a sparse open cluster, NGC 1647 (Melotte 26). This sparse cluster is easier to identify with the low power of binoculars than with a telescope, which may spread it out so much that it is difficult to discern as a cluster. With 10x50 binoculars, expect to see at least eight or nine stars against a soft background glow. Being able to resolve this many stars is actually quite impressive, given its enormous distance (1,790 ly). NGC 1647 is more than twice as big as the Hyades! A challenge: one of the stars that you can resolve is actually a close double star; can you identify which one?

We'll just swap the digits around to select our last object in this set, yet another open cluster – or is it? Recent observations suggest that NGC 1746 is most likely just a random, line-of-sight association of stars. You'll find it half way between NGC 1647 and El Nath. This lovely grouping is even looser than NGC 1647 – too much magnification, and you won't be able to distinguish it from the background stars. You should be able to resolve at least twenty stars with direct vision, and see some attractive chains and loops.

Chapter 15.2 – December: A Whale of a Time

Cetus, the Whale, never rises far above the horizon in the UK, but on winter evenings, it is about as well-placed as it gets, and has some lovely objects if you have a reasonably good southern aspect.

We'll begin with a star that is the prototype of an entire class of objects, the Mira variable stars. Mira (o Cet), whose name derives from the Latin 'mirabilis' (= 'miraculous'), brightens from mag 9.3 to about 3.4 (although it sometimes gets two magnitudes brighter than this) every 332 days (approx. eleven months). Its 2018 brightness is due in November/December, and will be approx. a month earlier on each subsequent year. Mira variables are slowly throbbing red giant stars in their late stages of evolution. As they expand, their surface cools to about 1400 K, so more of their luminosity is radiated in the infra-red part of the spectrum. This means they radiate less in the visible spectrum and so, to our eyes, they become dimmer. As they contract, they warm again, and so they emit a greater proportion of their energy as visible light, and they brighten.

Imagine that a line joining Deneb Algenubi (η Cet) and θ Cet is the base of an isosceles triangle and 1.5° NW of it you will find the 5th mag 37 Cet that lies at the third apex. 37 Cet is nominally very easy to split at x15, as its components are 49 arcsec apart, but the

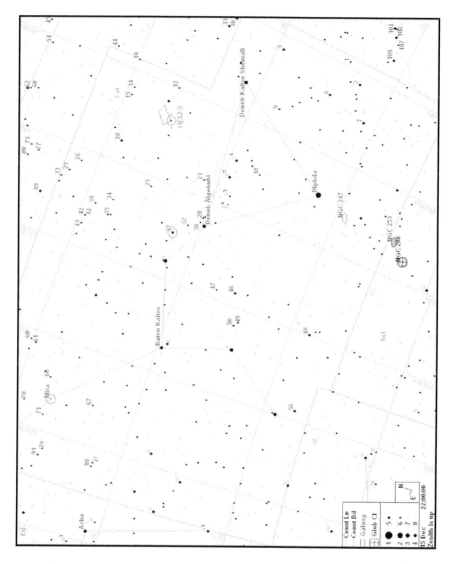

magnitude of the primary is more than twelve times that of the 8th mag companion. This difference in brightness can make it tricky to see the fainter star, so you may want to mount your binoculars for stability.

Identify 13 Cet and, from there, pan 1.5° to the ESE, where you will find HD3807. It is the southern corner of a not-quite-rectangular quadrilateral with sides about 1° long. The quadrilateral is brighter in the S quadrant, which seems full of stars, but seems to be devoid of stars in its N quadrant. HD3807 is also known as HJ323, a double star that was discovered by Sir John Herschel, son of William (hence he "HJ" – "Herschel, John" – in its alternate designation). The brighter component, which shines at mag 5.9, is nearly ten times brighter than the white secondary (mag 8.4), which lies 63 arcsec away in the direction of 13 Cet.

We'll finish off with a couple of more difficult objects. From further south, they are quite easy, but at the altitude at which they culminate in the UK, atmospheric extinction knocks about a magnitude off their brightness.

Just over 4° S of Diphda (β Cet) you'll see a little (¾° sides) triangle of 5th mag stars. About 3° to the S of this triangle, the Silver Coin Galaxy (NGC 253) appears as an elongated patch of soft light. Its long axis is about half the apparent diameter of the Moon, and you should be able to see that it has a brighter core. Despite its southerly declination, this is a relatively easy object for small binoculars, with the proviso that you have a decent southern horizon. Late autumn/early winter is the best time of year to observe it in the evening from Britain.

While you were looking for NGC 253, you probably notice that there is a smaller, softer, circular glow in the same field of view, a little less than 2° to the SE (the direction of α Scl) of the galaxy. This is the 8th mag globular cluster, NGC 288. Like the Silver Coin, it is a relatively easy target as long as you have a good southern horizon. Try averted vision and you should see that it expands to about half the width of NGC 253. If you ever wondered where the South Pole of the Milky Way is, You're close to it: it lies 36 arcmin from NGC288, towards the 7th mag star (HIP 3788) that you'll see 2° to the SSW.

Appendix A – Star magnitudes in Collinder 399

Magnitudes are in "deci-mags", i.e. 906 means mag 9.06.

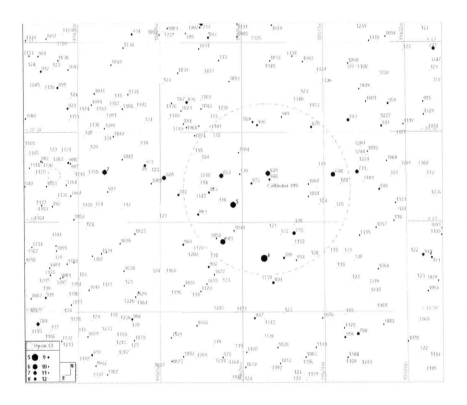

Appendix B: Binocular Designation Letters

There is a potentially bewildering array of letters that manufacturers use to give further information about binoculars in addition to the magnification and aperture. There is not an industry-wide standard of these, but here are most of the ones in recent and current use:

A: Armoured, usually with rubber (see GA, below).

AG: Silver coating on reflective surfaces of roof prisms (from argentum, the Latin for silver, whose chemical symbol is similar: Ag)

B: Depends on context and/or source of binoculars.

(a) Usual American, Chinese and Japanese usage. A Porro-prism binocular with each optical tube of one-piece construction (Bausch & Lomb or American style)

(b) Usual European usage: Long eye relief, suitable for spectacle-wearers (from brille, the German for spectacles).

C: Depends on context.

(a) Compact binocular (usually a small roof-prism binocular).

(b) Coated optics.

CF: Centre focus. Usually combined with another letter, e.g. BCF: Bausch & Lomb style centre focus.

D: Roof prism binocular (from dach, the German for roof).

F: Flat-field technology

FC: Fully Coated optics.

FL: Fluorite lenses.

FMC: Fully Multi-coated optics.

GA: Rubber Armoured (from gummi, the German for rubber)

H: H-body roof prism binocular.

IF: individual focusing eyepieces. Usually combined with another letter, e.g. ZIF: Zeiss style individual focus.

IS. Image stabilised.

MC: Multi-coated optics.

MCF: Mini Centre-focus ("delta" Porro-prism binoculars, with the objectives closer together than the eyepieces)

N: Nitrogen-filled

P or PC: Phase-corrected prism coatings (on roof-prisms).

P*: Proprietary (to Zeiss) phase coatings

SMC: Fully Multi-coated optics.

T*: Proprietary (to Zeiss) anti-reflective coatings

W, WA, or WW: Wide angle.

WP: Waterproof.

Z: Porro-prism binocular with each optical tube of two-piece construction, with the objective tube screwing into the prism housing (Zeiss or European style).

Appendix C: Care and Maintenance

"The best way to clean any optical surface is to not let it get dirty in the first place."

When you are storing your binoculars, replace all the lens caps and keep the binoculars in their cases. This will both protect the lenses and prevent accumulation of dust, which could otherwise easily enter exposed mechanisms.

When you are using hand-held binoculars and they are slung around your neck, use a rainguard-type of eyepiece cover to protect the eyepieces from descending debris. If it is tethered to the strap, using a rainguard in this way soon becomes instinctive.

After an observing session, when you bring the binoculars indoors, do not cap the lenses, but lay the binoculars horizontally and leave them for a few hours until any condensation has completely disappeared. Then cap them and put them in their case.

A little dust on the lenses will not have a noticeable effect on the image. If it bothers you, you can blow it off with a puffer-brush. However, there are some things that should be removed immediately you notice them. The first is fingerprints and eyelash-oil. Both of these contain chemicals that can etch the lens coatings. The other is pollen. Pollen grains can be extremely abrasive, and they can also adhere to surfaces they land on.

When you need to clean the lenses, the safest and most effective way is to use a proprietary lens cleaning system like Baader's Optical Wonder™ or Opticron's Residual Oil Remover™, which not only remove most pollens as well as fatty deposits from fingers and eyelashes, but also help to remove bacteria and fungi and to prevent their return. Use these fluids exactly in accordance with the instructions.

Lastly even if your binoculars are specified as being waterproof, it is a good idea to keep a sachet of desiccant in your binocular case. This not only guards against damage from moisture, but also makes the case an unwelcome environment for any bugs that my otherwise decide to reside in it.

Appendix D: Measuring pupil size

There is enormous variation in the size of the pupils of adults. A 2011 study of 253 18 to 80-year-olds found a range of 3.5 mm to 8.8 mm across the sample and, although the trend was for pupils to decrease with age, age was not an absolute predictor, because the decade age cohorts all had overlapping ranges.

The lesson from this is that there is no hard and fast rule. If you want to know the size of your dark-adapted pupils, you must measure them, at least approximately.

This is easily done by standing in front of a mirror at night, waiting 30 seconds or so for your pupils to dark adapt, hold a ruler under one eye, in the same vertical plane as your pupil, photograph the eye in the mirror (use flash, but not anti-red-eye pre-flash), then use image manipulation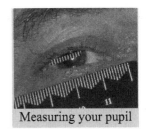

Measuring your pupil

software to move a section of the ruler over the diameter of the pupil.

Ideally, you will then acquire binoculars whose exit pupils are no bigger than your eye's pupil.

Appendix E: Testing Binoculars

There are tests that you can do in daylight, either in the store, or at home if you bought by mail order.

If you are satisfied with the binocular you have evaluated, buy that one, not an untested one in an unopened box!

When conducting these tests, it is important to remember that no binocular will be perfect; essentially these will help you to compare binoculars. Also, bear in mind that anything that is merely an irritant during daylight may become intolerable at night.

- Reject any binoculars that have zoom, ruby coatings or are "zip" focus.

- Give them a good shake. Reject any that have any internal noise.

- If they are of the type where the objective barrels screw into the prism housing, check that they are not cross-threaded. The tell-tale sign is an uneven gap between the objective tube and the prism housing.

- With a bright light behind you, look into the objective end of the binocular for signs of dust or other foreign matter.

- Check that the focus mechanism and right eyepiece dioptre adjustments are each smooth throughout their range, with no loose spots and no binding.

- Check that the central hinge is smooth throughout its range, with no loose spots and no binding.

- Check that the bridge connecting the eyepieces (in Porro-prism binoculars) does not rock under light or moderate pressure. This will cause the eyepieces to defocus in use.

- If you are going to use them with spectacles, verify that you can see the full field of view with the eye-cups folded/twisted down: the edge of the field of view should be a sharp black circle.

- Check fold-down rubber eyecups for splitting. Also check that they are not concealing any damage to the eyepiece tubes.

- Check twist-up eyecups for smoothness of operation and that they lock in place when extended or in steps (as appropriate). Also check that they are not concealing any damage to the eyepiece tubes.

- Check that your interpupillary distance (IPD) can be accommodated by the binoculars. In particular, if you have close-set eyes and/or a wide nose bridge, ensure that the eyepieces do not uncomfortably squeeze your nose. This is a particular problem with some wide-angle eyepieces.

- Hold them at arm's length and look into the eyepieces. Is there a round circle of light, or is it diamond-shaped? If the latter, the prisms are under-sized. If the cut-off bits are blue-grey, then it's full sized prisms, but they are low index glass. This need not be a problem for binoculars used in good light or with narrow fields of view.

- Some budget binoculars effectively stop down the aperture internally; this can make a 10x50 effectively a 10x42, or a 15x70 effectively a 15x63. You can check for this by shining a bright light into the eyepiece from a distance of 15cm or more. The circle of light emerging from the objective lens is the effective aperture of the binocular.

- Take them to the door of the shop and look at a distant, high-contrast, target (e.g. TV antenna against a bright sky), and focus it as best you can. Does it snap to a good focus or is there a small range of "nearly there"?

- How bad is the colour fringing across the field of view? (there will be some except in very high-quality binoculars).

- How far out to the edge of the field of view is the image sharp? (it will break down towards the edge except in very high-quality binoculars).

- Close your eyes while looking at something straight across the field of view (e.g. a roof ridge). Open them again. Were the images initially vertically displaced from each other? If so, reject the binoculars. (This is "step", and it should not be perceptible at all -- it leads to eye strain and headache.)

- Repeat the above with something vertical (e.g. antenna mast) to test for lateral displacement (convergence or divergence). A tiny amount is tolerable, but it's better to start off with none at all. (Eye strain again.)

- Move your vertical object to the edge of the field of view. It should curve slightly inwards at the middle. This is "pincushion distortion". A small amount is desirable, but it shouldn't be obtrusive near the centre of the field of view.

- Back inside the shop, target something small and bright, like a halogen ceiling light or the LED flashlight on a mobile phone. As you move it out of the field of view do you get any false/ghost images? Do you get bits of light from it even when it is out of the field of view?

There are some additional tests you should do on used binoculars:

- Give the binocular a thorough visual inspection for evidence of repair or tampering. If there is any, try to find out what has been done.

- Wherever possible, check for damage under eyepiece rubbers (by palpation if they cannot easily be lifted).

- Inspect external optical surfaces for scratches. This is best done viewing the surface at a glancing angle under bright light, which may reveal fine sleeks that result from improper cleaning. Eyepieces are particularly prone to this, since they tend to accumulate more debris, which is often wiped off with a tee-shirt hem or similar.

- Look for evidence of failing optical adhesive between lens elements. This may have a milky appearance in patches, and often starts at the edge of the lens. It is usually prohibitively expensive to correct.

- Look for signs of fungal growth. This is most likely to be seen on the edges of lenses, where it can penetrate between the lens elements. If it has penetrated, it is expensive to correct.

- Compare the view through both sides. If there is internal optical damage, it is usually not symmetrical, so one side of the binocular may have a different apparent colour, a different clarity, or a different amount of light scatter.

Object Index

The index refers to the chapters in which the objects are mentioned. The Greek (Bayer) letter entries follow the Roman alphabet (A-Z) entries.

ABOUT THE AUTHOR

Stephen (Steve) Tonkin lives in England on the edge of the New Forest, but was raised under the dark skies of tropical Africa, which fed his childhood passion for astronomy. He spends most of his time doing astronomical outreach with the societies of which he is a member, independently, and in his role as a STEM ambassador. He has taught astronomy to people of all ages for more than 30 years and has authored many articles and several books on practical aspects of astronomy. He writes equipment reviews and a monthly Binocular Tour column for BBC Sky at Night magazine.

Printed in Great Britain
by Amazon